Requiem

for a

Dying World

John Omaha

Daishe Bisghwaldo Press

First Edition

Library of Congress Cataloging-in-Publication Data

Names: John Omaha, 1938- author
    Title: Requiem for a Dying World
    Description: Paperback edition/ Santa Rosa, CA: Daishe Bisghwaldo
    (2019)
    ISBN 9781797499949

Prologue

You are not going to like reading this book. I might as well tell you now. You will struggle with reading it. It's not the concepts that are threatening, it's the emotions that will come up for you. Fear and shame and anguish. You will put it down and try to forget about it, but you will always return to it. It will call to you. As painful as it is to read this book, you know you have to. You know you must. You fear for the future, because suddenly the world you have known is unraveling. You try not to pay attention, but somehow you know that the planet is warming and the oceans are rising and the hurricanes are getting worse. You know this even though you try not to pay attention to the news. The subject of this book is death, the death of our world and almost everything in it, as well as the death of our way of life, and worst of all your own death. You are not going to like reading this book.

And yet, death has fascinated you all of your life, ever since you first learned about it. Death fascinates you. Why else would you go to the doctor if not to have him tell you how and approximately when you're going to die. This book summarizes the data that explain how the world is dying. More than that, it tells you what it is about us humans that is causing us to murder the beautiful world we live in. You will read this book because as you understand finally why we're literally burning down our own house with ourselves and families in it, you may experience some hope that the situation could change. It's unlikely, but it could happen. You will realize you can only change your own behavior. You can only change your own personality. In this book you will find a program for that change. Whether the world saves itself or not, your life will be better, no matter what happens, if you practice the principles presented here.

Introduction

The organism lay dying. Its systems and organs failing, it gasped for breath. Mechanisms that once kept the organism's metabolism stable no longer interacted, and its organ systems that had kept the massive being functioning harmoniously lost connection with each other. Individual cells and groups of cells expired. A great Contagion spread throughout the body infecting every system, every cell, eating away at the organism's ability to recover, to continue to nourish itself. The organism knew it was dying, at least some of the cells that stored awareness knew it was dying. The organism felt sad, at least those cells that stored emotion knew it felt sad, felt the sadness. Those same cells felt helpless and hopeless. The organism could not cure itself. The cancerous Contagion had advanced too far, had infected too many systems and organs and cells in the body. The organism could see itself as if in a mirror, could see the wasting, the deep lines on its surface, its skin. Balance was lost. The organism knew it was dying.

Many cells lived in denial. They swarmed through the dying organism muttering platitudes. "You're getting better." "You'll beat this." "You are strong." "Your doctors are working night and day to find a cure." Other cells prayed. "God did not create you to just let you die." "God is almighty. He will not desert you." "God loves you." "You are pre-eminent among all of God's creation." The organism knew better. It could feel itself failing. It could feel itself dying. The organism stopped listening to the deniers and the religious.

The organism comprised many types of cells, a wide variety of cells of all kinds. When the organism had been healthy there was a vast multiplicity of types of cells. Some cells fed the organism and all its different types of cells. These feeders made energy compounds that were the basis of life for the vast multiplicity. Some cells were predators, and their job was to kill and eat other types. In this way a balance was maintained among the types of cells. Some cells swarmed in the circulatory system, the enormous watery domain that bathed the organism. Other cells flew in the airy passageways that moved life-giving oxygen around the organism. The cells comprising the organism had lived in harmony for eons, but now the harmony was disrupted. The cells that made up the organism were dying. The specialized cells that held the organism's ability to know what was happening to it were disappearing because the cancerous Contagion encroached on every domain.

The world we have lived in is dying. The world is the organism whose death is described in the above scenario. Earth's oceans and rivers are the circulatory system. Earth's atmosphere is the breath of the organism. We humans and all the other organisms we share this planet with are the cells the scenario talks about. We humans are so caught up in the daily activities of just surviving that we have little time to appreciate the magnificent organism our world is. We are unaware of our intimate relationship with our world. This book will teach the skills of awareness and will provide exercises to deepen readers' appreciation of what our environment once was. The book transmits a comprehensive world view, one that encompasses the entire history of our planet from its birth through our evolution from the very substance of the planet to the death of our world, a death process that is well underway. Some of us are fortunate enough to have known the world when it was still somewhat pristine. The book offers a spiritual experience of the world as it once was before the Contagion. Readers will learn about the great Contagion, the cancer that has spread throughout the body, the Contagion that is killing the organism, the world we live in. Most importantly the book will explain the Contagion, what it is, what caused it, and how it is spread. The Contagion is a disorder of thought, emotion, behavior and spirit. Readers will gain a deep psychological understanding of the Contagion and how it originated in the conditions of our early evolution. As readers arrive at an understanding of the Contagion they will realize why humans have not raised the Contagion to awareness and why they have been unable to stop its spread. Whether we humans are able to stop the spread of the Contagion or not, readers will experience a grief process for the loss of the comfortable ways of the Contagion, ways that are disappearing for ever.

As the Contagion spreads and the world's death process accelerates, the established ways of living that we've accepted as a given are changing. Many of these ways of living were made possible by the Contagion itself. We humans did not realize the price we would eventually have to pay for the vast military, financial, and manufacturing systems we created. Humans did not foresee the consequences of our actions. The Contagion results from a spiritual wound, the dissociation of humans from an intimate relationship with our world. The death of our world is already provoking spiritual crises in America and around the world. Spiritual crises require spiritual solutions, and this book is a very practical spiritual treatise, because it provides spiritual solutions to readers dealing with the imminent destruction of the accustomed patterns of behavior, the usual ways of thinking and relating, and ultimately the destruction of the physical planet itself, our home world, Earth. As these established ways and accustomed environments collapse, it will be easy for people to collapse as well and to fall into disorganized thinking and behaving. This book will help you to stay sane, because it will provide you with a structure to understand and make sense of these profound transformations. Having a structure anchors people and helps them stay calm. It is unlikely that this book will change the course of the world's collapse, but if enough people read it, if enough people were to heed its message, perhaps the arc of the sixth extinction could be altered.

This is not the first time the world has died. Five times before across the five billion year life span of Earth life on earth died back. But for the first time, one of the life forms on the planet is causing the death of the world. Previously the causes were natural, most recently 65 million years ago, an asteroid hit Earth and in the aftermath the sun was blocked out by smoke and plants and animals died off. I use the word "world" because "planet" is too sterile. Pluto is a planet, and Pluto has no life forms. Earth is a world because it is more than the physical planet. "World" encompasses the physical planet and the organisms that inhabit it—every living thing; what the Lakota speak of as *Mitakuye Oyasin*—and the meanings and histories and spirits and ancestors of everything on the planet.

This book celebrates the life and acknowledges the death process of our world. Ritualizes the death process. A requiem is a mass said for the dead or any ceremony that is like a ceremony—a funeral, a wake—for the dead. When, near the end of the nineteenth century, the native people realized their culture was gone, they created the Ghost Dance. Ghost Dance was a ceremony for the dead. For the death of millions of native people and for the death of their culture, a way of being. Ghost Dance was a requiem.This book offers concepts that are vital to understanding the necessity of a requiem. These include a discussion of awareness, because we cannot truly ritually celebrate requiem unless we are more fully self aware and self conscious. A fuller awareness is required to break through denial and comprehend and accept the loss we are experiencing. A fuller awareness will help us accept with calmness the end of the world.

The book describes the evidence that the world is moribund. It lies dying.

The book describes the Contagion, the cancer of thought, emotion, attitude, soul, and behavior that has brought us to the hospice bed of our world, and it describes how the Contagion began and why it has been passed on and why it has proved impossible to stop. Readers will learn the awareness skills that are necessary to purify the self of the Contagion, which is a necessary step to acceptance that the world and the ways of life we have known are dying.

The book presents a ritual for celebrating the death of our world. Celebrating the loss of what was majestic and unique and meaningful.

Chapter One

Body/Mind/World

The world is dying. Many people do not understand that the world is dying. We humans have a great capacity to separate our experience into compartments that don't interact, and so it is easier for us to know that the arctic ice is melting and at the same time not feel it as a personal experience, one that will certainly affect us. In this way we humans can go on with our lives without changing how we live. Humans will never change until they experience situations personally. One can know that the oceans are rising, but until our own feet are wet, we won't take action. In order to personalize the message of this book, we begin by creating a concept that unites the whole of our experience with the context, the world in which we exist. We are an integral part of the world, not a split off fragment of it. The unifying concept is body/mind/world. The world is dying, and in order to understand that fact and how it came to be, we need to embed our body/mind in the world of which we are an integral part. The world is dying because the body/mind/world is dying.

The fragmentation of the body/mind/world had been underway for centuries when the French philosopher Rene Descartes wrote in 1637 "I think, therefore I am." His statement established fragmentation in western philosophy. He equated existence with thought. His dictum abstracted mind from body/mind/world. Descartes affirmed a theme that had been developing in the evolution of human consciousness for centuries: a disconnection between mind and body as well as a disconnection from the world in which the mind and body exist. Humans were losing touch with the embodied mind and with the context of the world of which the body and mind are integral parts. Three hundred-fifty years would elapse before psychologists began to reunite the body and mind. Psychologists observed that childhood misfortune, adversity, and trauma express later in life through physical symptoms.

Psychologists and physicians have become aware that the immune system can be affected by negative thoughts. Researchers began to realize that emotions are felt as bodily sensations. The only way humans know they are feeling anger, for example, is because they experience tenseness somewhere in the body, often in the neck, jaw, or shoulders. They may not be aware of the somatic experience, but that does not make it less true. Emotional experience embodies the mind. The body/mind was fragmented by a destructive belief that disembodied the mind and untethered if from its physical connection. The first step toward wholeness began when psychologists realized the mind is embodied. Mind cannot be separated from the body in which mind exists. Mind is a bodily function.

The next step toward wholeness consisted of setting the embodied mind in the environment of which it is an integral component. Fragmentation of the body/mind/world is a European phenomenon. Indigenous cultures, especially the native cultures that inhabited North America did not fragment their world. The people lived in oneness with the physical and spiritual environment. The concept of Mitakuye Oyasin united the physical, emotional, and spiritual world seamlessly. Mitakuye Oyasin means All My Relations and proclaims that each individual is an integral part of a whole that comprises every living thing, both animate and inanimate, every thing that ever lived, that lives now, and that will live one day. When Europeans arrived in the western hemisphere in the late fifteenth century, they brought the cancer of fragmentation of the world view, and the history of the world since then is a story of the clash between these two fundamentally different cultural systems: the European world view of fragmentation and the Indigenous world view of oneness.

While the Contagion of fragmentation flourished, a unifying concept of body/mind/world was growing among western thinkers as a radical alternative to the cancer of fragmentation. Having situated the mind in the body, these observers now created the emerging field of ecopsychology. Ecopsychology states that the body/mind has been conditioned by tens of thousands of years of evolution in the context of the physical world. The current social environment is believed to affect a mind molded by ancient environmental and social forces. The concept of body/mind/world proposed here deepens and expands ecopsychology's world view. Ecopsychology proposes that we humans have an emotional affinity with the natural world, and this appears true. Subsequent chapters will demonstrate how the conditions of our early experience as a species determined the development of attitudes, thoughts, emotional responding, and behaviors that became the destructive Contagion that is causing the world's death. The concept of body/mind/world states that the global warming; drought; destructive weather patterns; rising oceans; pollution of air, water, and land manifest in the physical world a dysfunction of the mind expressed through the body. In the next paragraphs we'll explore the meaning of the body/mind/world concept more fully.

Body/mind/world unites the rivers of the world with the rivers or vessels carrying blood through our veins. The rivers of the world flow through my veins, flow through your veins. Allow your mind to open to this concept. It is more than a mere metaphor. My aorta is a Ganges running north and south through my body. Your aorta is a Ganges running north and south through your body. Mind can comprehend that our aortas are a Ganges running north and south through our bodies. Notice how body/mind/world unites our individual bodies with our individual minds and with the context of the physical world. The Ganges is a major river, a holy river of India. Bearing water down from the Himalayas, the Ganges empties into the Bay of Bengal. The Ganges of my aorta carries blood throughout my body. The Ganges is holy. My aorta is holy, and so is yours. Mind brings the holy, and body contributes the physical sensations of holiness, and body/mind/world unites every living thing in a perfect embodiment of the holy.

Body/mind/world unites us with all beings, because all beings—from bacteria, to trees, to whales—have a circulatory system. The planet has a complex circulatory system that includes the oceans, atmosphere, sun to power the system by evaporating water, and rain and rivers to distribute the water. The holy waters of the Ganges flow through the aortas and circulatory systems of every living thing. Body/mind/world creates oneness among all beings whether animate like us and whales and wolves and hippopotamus or inanimate like the great planetary circulatory system. Whether we are aware of it or not, whether we acknowledge it or not, we are all inseparable from the body/mind/world. Mind, influenced by the Contagion we'll learn about presently, can tell us we are separate from body/mind/world, but that is a lie, a lie that many believe, or are unaware of, or at least do not question.

Breath also unites every living thing. We humans and multi-celled organisms breathe in oxygen and breathe out carbon dioxide. Breath is considered sacred in some spiritual practices, and life depends on the oxygen we breathe in. Our respiration is matched by the photosynthesis and carbon fixation of the plants including the great and small algae of the ocean. We breathe in oxygen and they exhale oxygen. We exhale carbon dioxide and the plants breathe it in. Sacred and holy breath exchanged among all living beings. Again, body/mind/world unites us. What I will call holopsychology elevates this concept to awareness.

We live in a time when we have forgotten the body/mind/world. The people believe they are separate from the world. The Contagion tells people they are separate. They have forgotten the womb of their mother as they have forgotten the womb of our mother the earth. They have forgotten the pulsing of blood in their mother's aorta, the first sound they heard, the whooshing of blood through her heart. Once they have emerged into the world at birth they are socialized out of connection with the blood-filled aortas and water-filled Ganges. Their culture fills their minds with Contagion thoughts, and Contagion emotions, and they lose oneness with their Columbias and Mississippis and Seines and Amazons and Ganges. They lose oneness with the breath that unites us all in the great body/mind/world.

The Contagion comprises a belief system that contaminates the body/mind/world. The contaminating thoughts, beliefs, attitudes, and emotions manifest as pollution of the circulatory systems of the physical world as well as the biological world. Humans have polluted their own aortas with toxics and they have polluted their Ganges and their Seines and their Columbias, their Mississippis and Amazons. They do not understand, comprehend, or accept the womb of our mother the earth, the mother of us all. The body/mind/world is lost to them. They are lost. They wander desperately. They want the connection, yearn for it, yet they do not know how to reclaim what once was theirs. Their religion preaches dominionism and drives them further from connection, oneness, and majesty of the body/mind/world.

Driven insane by the Contagion teachings, humans dump the waste of their extractive practices into the aortas of their body/mind/world. The body/mind/world rebels. Revolts. A great cancer has taken over the body/mind/world. The Contagion is a cancer. A cancer of the body, of the mind, of the world. Greed, money, self-absorption, violence, mendacity, and disconnection make up the cancer and enable its metastasis. The Contagion metastasizes throughout the body/mind/world. The Congresses world-wide are infected. The Executive branches in America and many other countries are infected. Corporations embody the Contagion and are primary agents of its spread. Militias of infected human bodies and minds patrol the streets intimidating citizens who know better, but are frightened into silence. Violence promotes metastasis. Cars ramming protesters who do know the meaning of body/mind/world and killing them are metastases of the Contagion, the cancer that spreads through the body/mind/world. Contagion! A mighty voice screams "Contagion" and Contagion spreads throughout the body/mind/world. The mind is infected. The Contagion has metastasized to the and mind created false enemies and screams "Kill the Jews" and "Kill the faggots" and "Kill the Muslims," when in fact the true enemy is the Contagion. Contagion murders in churches, and it murders in schools just as it murders rivers and lakes and land and air. Contagion has driven humans insane.

The Contagion has spread out of the humans' minds and bodies and now infects the world itself. Contagion disrupts the functioning of the physical world as it has disrupted the functioning of the human beings. Jet streams in the atmosphere are disordered, crossing the equator for the first time. Storm bombs suddenly blast out of superheated air and dump inches of rain and tennis ball-sized hail. Droughts parch the fields. A massive volcano may soon erupt in Yellowstone. Glaciers, the Greenland ice sheet, and Arctic and Antarctic ice are melting. The planet is warming, and oceans are rising. All of these phenomena demonstrate that the Contagion infects the world. The Contagion of human bodies and minds now explodes into the world itself.

In the human body, when a cancer reaches the pancreas, it is a death sentence. Humans usually die within a couple of weeks of being diagnosed with pancreatic cancer. Unfortunately, the Contagion has reached the pancreas of the body/mind/world. Drought is drying up the fields and basic food crops are failing. Famine is spreading in Africa endangering the lives of 20 million people in Somalia, South Sudan, Yemen, and Nigeria. The famine in Africa is a harbinger of what will come for humans world-wide. Pakistan and India are at risk. America itself is just one drought year away from a similar fate.

Requiem for a Dying World proposes to celebrate and honor and remember the world that once was and is now disappearing as the Contagion advances. Ritual is the process we humans have devised to support each other when there is a great loss. Our Contagion ways that have brought us a false prosperity and the illusion of a stable, healthy existence have also brought us the cancer of our body/mind/world that is killing our bodies and our world. We perform a requiem for the false prosperity. We perform a requiem for the beliefs of unlimited growth, of a money economy, for the self-absorbed belief that we had dominion over nature, for the narcissistic belief that we could separate ourselves from the natural world without consequences. We perform a requiem for what could have been if only we had been wiser and less influenced by the Contagion that was created eons ago and that we all inherited from our ancestors.

Chapter Two

Awareness

In order to understand Requiem we must understand awareness. Requiem arises from a profound awareness of our relationship to every living thing. We begin our Requiem by developing an understanding of awareness.

What is awareness?

I am aware, therefore I am alive. Awareness manifests the ongoing biological activity of the body-mind. True awareness situates the mind in the body. It is so pervasive, such an integral aspect of daily life that we assume it, we do not devote much time to noticing it. Awareness includes a knowing that clenching my jaw signals I am feeling anger. Awareness comprises a knowing that looking down with my eyes and a collapsing in my chest tell me I'm feeling shame or that tears in my eyes and a heaviness in my chest are the signs that I'm feeling sadness. A warmth in my heart means I'm feeling attachment affect. A rapid heart rate and shortness of breath accompanied by sweating palms tells me I'm feeling fear. The body is integral to awareness. The mind receives these somatic signals and makes sense of them. The mind is essential to awareness. Awareness begins during gestation and grows across the lifespan. Awareness is co-extensive with life. I am alive because I am aware.

Awareness ceases at death and we fear death because it is the permanent loss of awareness. Awareness is the foundation of this book. Awareness is an experience. It is fundamental to grasping the message of the book. Let us begin with a focusing exercise to raise the experience of awareness to awareness.

Sit quietly. Let all intrusions disappear. Notice your body at rest. Notice your breathing. Notice the expansion of your lungs as you breathe in. Notice the collapse of your chest as you exhale. Let the end of one breath become the beginning of the next. Let your noticing sink down inside you. Realize that your noticing is your awareness. Experience yourself concentrating your inward focus. Your inward focus is your awareness. With all external and internal distractions evaporated, notice your pure body-mind. Condensed. Focused. Sit for a moment in this experience of body-mind awareness.

Now allow your awareness to move outside of you. Just a little bit at first. Become aware of the plants and trees in their seasonal mode. You know they are there. Maybe in your yard or the neighborhood or a park or the forest at the edge of your town or city. Just be aware of their presence. Be aware of the little birds, house sparrows, that you saw hopping from branch to branch earlier in the day. Crows. Pigeons. Be aware of them. Wherever you live, be aware of the animals. Raccoons. Cats. Maybe coyotes. You know they are there. Just become aware of their presence. You do not have to see them to know they are there. Let your expanding awareness encompass them.

Water. From where you sit, eyes gently closed, become aware of the water. A pond. A stream. Perhaps a river. Perhaps an underground aquifer. You've seen so many bodies of water in your life. They are still there. Let your awareness move out to them. And rocks and mountains. Sense the marble, the basalt, the serpentine. They are there. Your awareness moving outward encompasses the stone people.

Allow your expanding awareness to move out beyond where you are sitting, out through the layers of the sky, through the clouds, becoming aware of each as you go. Giving awareness to all. Out to the moon. To the planets. To Sol, our sun. And now, become aware of our own galaxy, of the local group of galaxies, of all the galaxies in the universe, giving awareness to them.

Finally, become aware of yourself. We humans are gifted with self awareness. We can be aware that we are aware. Bring this entire meditation together with the knowing that you are aware of all and of yourself being aware of all. Hold this. We humans come from the void, the emptiness, the unknowing. As we grow and learn we become more and more aware. We are the means by which the universe is aware of itself. We are the bearers of self awareness. Eventually we will return to the void, and our awareness and self awareness will slip away. Our prayerful intentions and our participation in ceremony and ritual all depend upon our awareness. Be aware. May you live in awareness all the days of your lives.

As you were participating in the awareness meditation, you may have noticed mental or emotional intrusions. You may have had judgments like "This is stupid," "This is out-there," "This is too hippy-dippy for me," "I have an appointment at ten this morning," "I can't be wasting time with this." You are becoming aware of your unconscious intruding. You may also have become aware of an unease, which is the emotional component of the intrusions. We will see how the mind affected by the cancer of fragmentation attempts to deny the wholeness of body/mind/awareness. Awareness of intrusions is a first step toward healing.

Awareness is experiencing. Awareness is noticing. Awareness is attending. Awareness is focusing. This simple focusing exercise or meditation introduces the experience of awareness. Awareness is so pervasive, so fundamental to our day-to-day experience of ourselves, so much a part of who we are, that we often do not notice it or think about it. We are often unaware that we are aware. We are often unaware that we are unaware of our unawareness.

Evolution of awareness

In the beginning there was no awareness; there was only the void. In the nothingness of the void, a cosmic explosion occurred, what cosmologists call the Big Bang. This occurred 13.5 billion years ago. This is one of the stories we tell ourselves about our origins. Pure energy came into existence. Where it came from is unknown, and that is the fundamental mystery of existence. Energy coalesced into matter. Hydrogen and helium were born. Some hydrogen and helium formed into stars held together by gravitation. Some of these stars were unstable and exploded in supernovas, and other elements emerged: carbon, oxygen, nitrogen, sulfur. These elements coalesced into globs some of which solidified and formed planets. In our solar system, the solid inner planets, Mercury, Venus, Earth, and Mars formed. Out beyond us the gas giants orbited. On our earth, as it cooled, powered by the energy of the sun, atoms united to form molecules. Molecules assembled into larger and larger structures. Over the millions of years, molecules emerged that could copy themselves. Other molecules emerged that could harvest the energy of the sun. With aeons of time, cellular organisms emerged, then colonial organisms that were aggregations of cells, then more complex organisms, bacteria, organisms that eat bacteria, organisms that eat the organisms that eat bacteria, a process that resulted in terrestrial animals.

As the organisms evolved, sensory systems emerged that promoted survival. Bacteria developed systems to sense sources of nourishment. Those sensing systems were the rudiments of awareness. As plants evolved from photosynthetic bacteria, sensing systems evolved to facilitate the plants turning towards the sun during the course of the day. Always driven by survival. What was successful survived, and what didn't work or didn't work as well, did not survive. Eventually more and more complex animals evolved. Simple chemical awareness evolved into eyes and ears and proprioception. At last, humans evolved and perhaps for the first time an organism that was aware of itself, aware of its awareness emerged. The thrust of evolution has been toward self awareness. We humans are the means by which the universe is aware of itself.

Awareness is what defines life. Being alive means being aware. Bacteria are chemically aware of their surroundings, and when they die, they cease to be chemically aware of their surroundings. Round worms, flat worms, and insects are aware of their surroundings, each with their own unique form of awareness. Bees sense the world in the ultraviolet range and move toward plants where they suck up nectar. Plants have a primitive awareness too. They receive information from the sun and from each other through pheromones. Plants are aware, but not conscious, of where the sun is, and through hydraulics and growth they turn toward the sun. Bears sense the environment through smell. Snakes register their world through chemical sensors. We all, animals and every living thing, inhabit a field of awareness. Life entails awareness. Awareness evolved as life evolved out of the planet's physical and chemical environment. Awareness cannot be separated from the world in which life evolved. Awareness unites the body/mind/world.

Human Awareness

Awareness is also known as consciousness. It is the capacity to experience the internal and external environments. Humans can be fully alert, completely aware of their surroundings, thoughts, feelings, and sensations. Human consciousness can be compromised to various degrees as in disorientation, delirium, loss of response to painful stimuli, coma, and death. We lose consciousness when we are anesthetized. We lose consciousness when we sleep.

We humans separate the contents of our consciousness into categories. A fundamental level is perceptual awareness. A newborn is perceptually aware of hunger as an uncomfortable emptiness. Later he or she will learn to name the state. We share perceptual awareness with other animals. Like the newborn and like all other animals, coyote is perceptually aware of hunger. Infants are perceptually aware of colors that later will be associated with names. Perceptual awareness is the raw experience of color, form, and sound. Sensations like itching, or tension, or nausea also comprise the experience of raw awareness. Along with the sensations that tell us what emotions we're experiencing these experiences are called qualia. We share this level of awareness with other animals across the evolutionary field. This raw level of consciousness is called perceptual consciousness.

A second category of consciousness, called "access consciousness," arises when the perceptions, sensations, emotions and other qualia are available for reflection, reasoning, and control of behavior. Remembering is access consciousness. For example, the color blue and the color yellow register in perceptual consciousness, and the thought that when they are combined they make green comprises access consciousness.

Awareness is an emergent function in humans which means that it develops over time as the newborn develops. As the infant develops, layers of experience enrich the perceptual consciousness phenomenon of hunger and it becomes access consciousness. The raw emptiness experience of the newborn assembles with parents' words and actions, and access consciousness develops to the point where the qualia of emptiness motivates the child's verbal behavior, and he or she says, "Mommy. Hungry." With the accrual of still more layers of experience and learning, self-awareness may emerge, and the child will be aware of itself as experiencing hunger. Eventually the child will learn to say, "I am hungry." Raw awareness, perceptual consciousness, is an inherent property of the organism at or before birth, and access consciousness develops from this basic awareness as the organism acquires more and more experience.

Development of Awareness in the Individual

Development of awareness in an individual born today reenacts the development of awareness in our species. For the first 4 years of life, the neural apparatus for creating narratives has not formed, and so we rarely have memories of those years. Brains are receiving information and storing it, but because we are not yet aware of it, it is called the unconscious. Even though we can rarely remember much of what happened to us in those first 4 years, the information that has been received and stored has an influence on our subsequent perceiving, thinking, feeling, and acting. Archeologists believe Homo sapiens first appeared 250,000 years ago. These ancient humans possessed perceptual awareness, but they did not yet have access awareness. Development of access awareness is a cultural phenomenon that requires language. For the first 200,000 years of our existence as a species, humans were apparently unconscious; we were not aware of self as a unique and separate entity. We absorbed vast amounts of knowledge about the environment, plants, other animals, and other human beings, and this awareness conditioned much of our behavior subsequently. Our unconscious, inherited in part from our ancient ancestors and received in larger part from our parents, affects our behavior and thinking today. The development of awareness in an individual reenacts the development of awareness in our species.

Psychologists have learned that humans can become aware of the effect of unconscious memories on current behavior, but to do so, the person must be able to be aware of their current behavior. The only way to know how what happened in childhood affects current behavior is to accurately be aware of current behavior. What is true for the individual human is also true for families of humans, for tribes of humans, for societies of humans, and for civilizations of humans. Awareness of current behavior is often difficult for people, because one of the lessons learned in childhood is called denial. Denial protects the person from accurate awareness of thoughts, behaviors, actions, and feelings, because the emotion of shame is often assembled with those thoughts, behaviors, actions, and feelings. Shame is an extremely powerful emotion. What gives shame its power is that it is physically painful to experience. The great dilemma we humans face in dealing with our individual problems and in dealing with our species' problems is that we have to raise our behavior to awareness to resolve the problems but we have very powerful mental mechanisms that prevent awareness of the behaviors that caused the problems. We will return to this theme again and again throughout the book.

An infant's early experience determines the structure of the infant's developing brain. The human brain is wonderfully plastic. Which neurons connect to which other neurons is determined by the infant's experience. An infant whose caregivers are attuned and appropriately responsive learns that a cry of hunger results in feeding. The infant learns this without knowing it is learning it, but the neurons connect nonetheless and a knowing forms. The infant is developing access awareness as it acquires the information that a cry will be followed by feeding. Infants whose parents are not sensitive, attuned, and responsive may cry and not receive food. That infant's unconscious acquires the knowing that its complaint will not always be attended to. The mental structure acquired in infancy affects behavior for the rest of the organism's life. As an adult, the infant whose cries of hunger were attended to will act as if the people in that adult's life will notice an emotional complaint and will attend to it. As an adult, the infant whose cries were not attended to will act in adulthood with the expectation that its cries will be ignored. An infant's experience determines how its brain organizes, and how the infant's brain organizes itself determines the nature of the adult's personality. A child's experience determines the quality of its awareness as an adult.

When a person's experience is rich enough in infancy and childhood and the brain develops fully enough, the highest level of access consciousness appears: awareness of self. Awareness of self includes introspection, the awareness of one's own actions and one's own thoughts and one's own emotions, the awareness of thinking in words and images. Inner speech capacity is also linked to self-awareness. Awareness of self includes the experience of one's own organism being the agent of perceiving, the agent of thinking, the agent of existing. Not every human achieves all of these stages. How far along the scale of awareness one progresses is experience-dependent. There is no fixed timeline for the development of higher levels of awareness. The fundamental level of awareness, perceptual consciousness, exists in all intact humans. A child born blind will not possess the perceptual consciousness of sight. Because the human brain is able to form new connections throughout the lifespan, higher levels of access consciousness can appear at any time that life experience provides the necessary conditions. For example, a person in recovery from addiction or alcoholism can become conscious of herself through working the steps of a 12-Step program. In addiction, the person acted without awareness of the meanness or self-centeredness of their behaviors. Through the steps or through psychotherapy, the recovering person examines his behaviors with the help of a sponsor or therapist and comes to acknowledge what he did to himself and to others, what he thought and what he felt. In the course of this work he develops an awareness of himself as the agent of his behaviors; he develops self aware self consciousness.

Awareness of self

Understanding Requiem will require a deep comprehension of awareness and especially awareness of self. This section will discuss what is known about the neurophysiology of awareness and self awareness. It will also cover the brain's ability to model the thinking, feeling, and behavior of others, what is known as "Theory of Mind." This section prepares us to discuss human awareness of reality, the actual state of the world. It also prepares us to discuss death.

Scientists who study the brain and its functioning often learn how the brain works by examining people who have damage to parts of their brains. Neuroscientists use assessments to determine what functions are lost when a particular part of the brain has been harmed. One assessment is the Frontal Systems Behavior Scale that measures symptoms associated with damage to frontal brain regions. Examples of symptoms associated with damage to frontal regions include mixing up a sequence of actions because of getting confused trying to do several things in a row. Speaking only when spoken to is another symptom of frontal brain damage. Another assessment of frontal damage is the Cognitive Failures Questionnaire that asks about the experience of forgetting appointments or of reading something and finding you haven't been thinking about it and have to read it again. The Measure of Empathic Tendency assesses how a patient feels when others are having strong emotion. The assessment asks the patient to rate how true a statement like this is for them: It makes me sad to see a lonely stranger in a group. Another statement is: I become nervous if others around me seem nervous. Brain damage can affect empathic awareness. Neuroscientists use questionnaires and assessments to determine how damage to specific regions of the brain affects self-awareness.

Awareness of one's own body is a central element of self-awareness. Damage to certain parts of the brain result in a lack of awareness of a part or all of one's body. Damage to other parts of the brain prevent a person from recognizing his or her own reflection as belonging to the self. This condition is called "mirrorsign," and it demonstrates a key aspect of self awareness, recognition of the self. Awareness of the self is a developmental phenomenon. It is measured in children using a mirror. Up to age 12 to 14 months, children placed in front of a mirror will react as if they were facing another child. They are not yet aware of themselves as separate, unique, individuals. Developmental psychologists use the "rouge test" to determine when self-awareness develops. A spot of rouge is placed on the child's nose and then the child is placed in front of a mirror. Up till about 16 months children do not notice the change in the image in the mirror. Beginning at about 18 months about half of children will demonstrate awareness that the image in the mirror is changed and they will touch the rouge or try and rub it off. They have become aware that the image in the mirror is an image of the self. By 20 months a majority of children will do so. Developmental psychologists infer that the rouge test demonstrates that awareness of self emerges between 16 and 24 months.

Self awareness depends on the interaction of many areas of the brain. The exact neuropsychology of self-awareness is a topic of active investigation. Damage to frontal lobes impairs self-awareness. Self-awareness is also linked to right-brain areas. Memory is an important component. The executive functions of strategy selection, planning, interference control and inhibition, goal maintenance, sustained self-monitoring, and attention are essential to self-awareness as well. Patients with damage to the right-side cortex often are unaware of how sick they are. Self-reference or self-monitoring is affected in a condition known as anosognosia, lack of awareness of how sick one is.

One reason humans were fit enough to survive the ardors of natural selection is the aspect of awareness called Theory of Mind. Self-awareness is the ability to monitor one's own mental state, that is one's thoughts and feelings, as if from a third-person perspective. Suppose I am experiencing tenseness in my jaw, neck, shoulders, and fists. Without self-awareness, these sensations are all that exist. With self-awareness, I would say, "I'm feeling tense. I am feeling angry." Theory of Mind allows me to observe another person, notice the tension in his jaw and fists and neck and conclude that he is feeling angry. Theory of Mind is the ability to model the experiences of others. The more aware and self-aware we are the better we can model the experience of others. For example, if I see my friend sitting alone, holding his head in his hands with the corners of his mouth turned down and his face sagging I will conclude, "My friend Bill is feeling sad." That is Theory of Mind. Like self-awareness, Theory of Mind is a right frontal lobe function. Self-awareness and Theory of Mind confer survival value because we can remember what we have done and how it worked out. Because we are self-aware, we can mentally travel in time. The games we play like chess or poker are based on mental time travel. The chess player calculates if he moves his knight to a certain position, he will capture an opponent's king in so many moves. That is mental time travel and it requires awareness and Theory of Mind. Other human skills depend on Theory of Mind including displaying empathy, consoling another, the difference between ignorance and knowing and between seeing and knowing. Chimpanzees do not possess the ability to see and know. For example, a chimp can see a researcher hide a banana, but that does not lead to knowing its location. A fundamental quality we humans possess that depends upon Theory of Mind is the ability to deceive and the ability to detect deception.

Awareness of deception depends on Theory of Mind. One must be aware of one's self in order to be aware of another's mental state or intention. Deception is an intention, the intention to make someone believe what is false, to delude someone, to cheat or swindle or double cross. The goal of deception is to gain access to a valuable resource, often money. Politics often seems to be motivated by deception, for example when a person runs for office making one claim and then does something else when elected. Financial gain is usually a motivating factor. Business can be motivated by monetary gain and can involve deception as when a used car salesman says the car he's selling has "another hundred thousand miles in it" but it really needs major repairs. Like self-awareness and Theory of Mind, awareness of deception is located in the cortex of the right hemisphere. Without awareness of deception, people can be cheated. They are unaware of the deceptions being perpetrated on them and they can lose valuable resources like money, retirement, insurance, social services, or a home.

Awareness and reality

Reality is terrifying. The map-making function of the human brain works to reduce the terror by placing order on the terrifying reality of the world. Raw reality is especially terrifying. Without the ordering function of the mind, humans feel a chaos that is like psychosis. As we progress from infancy to childhood we develop filters that make meaning out of disorganized reality and thus reduce the terror associated with it. The filters we form structure reality. For example babies form a filter that tells the baby the terrifying noise of a vacuum cleaner is associated with mom pushing the noise-maker around the room, and because mom is not terrified of it, the baby need not be. That filter structures baby's reality. The baby acquires information from observing mother. What was terrifying becomes benign. We humans have many filters that structure our reality. The filters are as individual as the people who have them.

Reality is essentially ambiguous. Reality has the meaning that we give to it. Reality has no inherent meaning. The history of humanity is the story of the emergence of meanings in succession over the life of our species. Meaning systems reduce the chaos that causes terror. Science is one such meaning. Science is a filter that gives meaning to reality and serves to decrease terror. Religion is also a meaning-making creation that serves to decrease terror. Religions—

Christianity, Judaism, Hinduism, Shinto, Islam—are all filters that give reality meaning and thereby decrease the terror of existence. Buddhism is another such meaning structure. So is shamanism and other earth-based spiritualities like Wicca and Humanity Rising, which is a synthesis of several meaning maintenance models. Meaning maintenance models order our experience of the world and reduce anxiety. We all live the meanings we make and through which we perceive and understand reality. From birth on humans are making mental maps of the world and how it works, and these maps put structure on our awareness. Our ability to construct the filters depends upon a unique human ability, the capacity to think about and reason about unobservable entities. Theory of Mind explains how we think about other human minds, which are unobservable. We also think about gods, which are unobservable. The central theme of this book, *Requiem for a Dying World*, depends on our human ability to think about people, cultures, ways of being and planetary processes and to make mental maps of where the world is headed.

Intention and awareness

In the world view of Humanity Rising we are one with the universe that we perceive. Every hydrogen atom in our bodies is 13.5 billion years old and originated in the Big Bang. The carbons, sulfurs, nitrogens and oxygens in our bodies and in the bodies of every living thing came from supernovae, stars that blew up and in the explosion these elements and all the naturally occurring elements of the periodic table were created. We are one with this marvelous universe. As evolution progressed, the elements organized into life forms. Simple ones at first, single-celled organisms. More complex and more complex still until one aeon mammals evolved from our reptilian ancestors. And then apes and then proto-humans and then, about 250,000 years ago the First Mother gave birth to the first Homo sapiens baby. We are the self-aware self conscious organisms. We appear to share self-awareness with elephants and whales and dolphins. Koko, the gorilla, developed a simple self-awareness as she learned sign language. We are the means by which the universe is aware of itself. Our awareness arises from the very universe of which we are a part and from which we evolved. Our thoughts, like our awareness, are part of the universe. Without the elements and atoms and molecules, there would be no awareness or self-awareness. Awareness and self-awareness are dependent upon the elements of which we are comprised. We are one with the universe. In addition to self-awareness, intention is a kind of thought that also arises from the complex assembly of atoms and molecules that we call brains.

Intention is a means for organizing experience in service of a goal: that which is intended. I will give an example from my spiritual practice. At the beginning of every sweat lodge ceremony as I am smudging each participant, I ask "What is your intention in coming to this sweat?" This is the moment at which the participant has the opportunity to organize his or her spiritual experience for the next several hours. We're going to be in the lodge for 4-5 hours on Saturday evening and another 4-5 on Sunday morning. We'll be purifying, journeying, praying, and transforming, and our intention organizes that experience. The most effective intentions are ones that are personal and involve only the participant. It's much more difficult to intend that others will change or that some event in the external world will happen like winning the lottery for example. "My intention is to become a better medicine man," is an example of a personal intention.

Intentions can also be formed by groups of people. Making change in the external world requires collective intention. Here is an example. In the Great Mystery Lodge of Humanity Rising we have formed collective intentions, that is we united our intentions toward a specific goal, in one case praying for rain here in Northern California during a protracted dry spell. A high pressure system off the coast out over the ocean can get stuck and block storms swirling up from the south or in from the vast Pacific. When this happens the "storm door" is closed and weather systems that would otherwise flow in are diverted to the north. In the lodge we united our intention and then sent our spirit bodies—what Don Juan Mattos as reported by Carlos Casteneda, called our *najuals*--out to merge with the high pressure system and then to intend that it move out away from the coast and into the open ocean. When we did this in the lodge, the storm door opened and it began to rain within a few hours. We take no credit. Our intervention worked that time. If all human beings formed the collective intention to free ourselves of the Contagion of greed-violence-usury-self absorption then peace would come on earth and we'd begin to repair the damage we have done and restore our planet to health.

Summarizing what we've uncovered so far in our exploration of awareness: Awareness is the ability to perceive, to feel, or to be conscious of events, objects, thoughts, emotions, or sensory patterns. In biological psychology, awareness is defined as a human's perception and cognitive reaction to a condition or event; awareness occurs on a continuum across living beings from the chemical awareness of bacteria through the primitive awareness we share with other animals to the self-awareness we appear to share with a few other mammalian species; awareness emerged in the course of evolution from the origin of the universe; awareness manifests from the functioning of several areas of our brains; awareness comprises a range of phenomena from perceptual awareness to process awareness to self-awareness; awareness develops over the individual's life span; awareness includes the ability to model another organism's behavior and mental state; and awareness can include consciousness of intentions about a future objective. Not all humans access the highest level of awareness. One's level of awareness is experience dependent. The richer one's experience from childhood on, the greater the level of awareness one will attain. Awareness and intention are interrelated. If one intends to increase one's awareness, perhaps through meditation or ritual, one can indeed enhance one's awareness.

Sharing awareness with All Our Relations

We humans exist in a complex interactive relationship that includes every dimension of our planetary environment. The greater one's awareness, the more likely one will grasp this concept. Whether one is aware of it or not does not change the fact that we are in an interactive relationship with everything in our environment. Native people conceptualized this interaction with the phrase *Mitakuyue Oyasin* which means "All Our Relations," and which refers to every living thing, whether animate or inanimate. We are related to the mineral part of of the environment through our very bones which are composed of the same calcium and phosphate found in rocks. The iron in the hemoglobin of our blood is the same as the earth's iron core. The planet nourishes us. We breathe in oxygen produced for us by photosynthesis in plants, and we exhale carbon dioxide that plants absorb to incorporate into sugars and into the carbon compounds of their woody bodies. We humans are intimately related to All Our Relations through cyclic exchanges of carbon, nitrogen, oxygen and other elements.

Many humans are not aware of these reciprocal exchanges with All Our Relations, but they exist nonetheless. Many humans are not aware of the meaning of these reciprocal exchanges. These reciprocal exchanges define our oneness with the world and everything in it. As we raise these interactions to awareness we may ask, what can we give back? For example, the biological economy of the planet depends on the oxygen produced by plant photosynthesis. Oxygen is a give-away of the plants in the reciprocal interchange among All Our Relations. What unique quality do we humans possess that we can give back? The answer is our self-aware self consciousness. Often people do not think of themselves as being in relationship with trees, insects, other animals, or with the mineral environment, with water, or with the sun. We have not raised these relationships to awareness. We take for granted so much of what All Our Relations provide us. Trying this simple exercise will help you change your awareness of All Our Relations.

Form the intention that you will raise your oneness with All Our Relations to awareness. Go to a park or a forest, somewhere that trees are abundant. Find a quiet place to sit and meditate. Empty your mind as well as you can. If thoughts arise, notice them, and let them go. Now, focus your awareness on a tree. It doesn't matter what kind of tree. Just sit with the tree and be aware of it. Notice its bark, its leaves, its structure. Allow your awareness of the tree to deepen and develop. Allow images and reflections related to the tree to surface and notice them. See the tree with your heart. If extraneous thoughts come up, let them go and return to the tree. As you sit with the tree become aware of your emerging relationship with the tree. Now you are sharing with the tree. It is giving you oxygen to breathe and you are giving it awareness of itself. Become attuned to the tree. Become sensitive to what the tree has to teach you. Appreciate the tree.

Doing this exercise may change you. You may find yourself noticing your relationship to sun, moon, air and sky, water, plants of all kinds, the mineral environment, insects, birds, and mammals. Notice how your thoughts and attitudes change. Notice the feelings that come up. Perhaps you will experience a lightness of expanding in your heart area the indicates you're feeling awe or marvel. Just notice the feelings that come up. You may find a more spiritual relationship emerging. Allow it. When we are aware of our interrelationship with Our Mother The Earth, then we will be respectful. When we are aware of our awareness of our interrelationship we will be caring and supportive. When we develop this awareness, we will realize our connections with All Our Relations and we will be less likely to abuse Our Mother the Earth.

Awareness of death

In his first teaching after achieving enlightenment, the Buddha formulated the Four Noble Truths for his students. The First Noble Truth states that "All life is suffering." The Buddha was referring to the suffering he saw among the people when he emerged for the first time as a young man into the world outside the protective walls of his father's castle. At that time in India he would have seen people suffering from leprosy, starvation, disease, and poverty. He had not known of any of this because his father had protected him. He was overwhelmed by the pain he saw, and he set out to discover a way to end the pain.

In his discourse to the *bhikkhus*, a name for student, the Buddha appears to be referring to the emotion that present-day psychologists who study emotions identify as distress-anguish. Distress-anguish is the emotion of too much. At the lower level if is distress; at the upper level it is anguish. The emotion states of grief and sadness lie on the continuum of distress-anguish.

In this discourse, the Buddha presents the Four Noble Truths. He presents the first as follows:

"Now this, *bhikkhus*, is the noble truth of suffering: Birth is suffering, aging is suffering, illness is suffering, death is suffering; union with what is displeasing is suffering; separation from what is pleasing is suffering; not to get what one wants is suffering; in brief, [whatever is subject to clinging is] suffering.

The Sanskrit word for suffering is d*ukkha*, which means incapable of satisfying. It refers to the unsatisfying nature and general insecurity of much of our human experience. *Dukkha* has been incorrectly translated as suffering. It is the opposite of *sukha*, which mean "pleasure." *Dukkha* is better translated as "pain." What the Buddha appeared to be saying is that all life experiences are painful, or to use a psychological concept, all life experiences cause distress-anguish."

The Buddha went on to propose that pain originates in craving, seeking delight in sensual behaviors. This is the second noble truth. As we will see, the Contagion that is destroying our world has its origin in craving for relief from the pain of existence. The third noble truth tells us how to cease suffering. In western terms, the Buddha appeared to be telling us that through acceptance, that is relinquishing craving, one can cease suffering. Giving up craving results in acceptance of reality as it is, and with that, pain ceases. Finally the Buddha proposed a healing process for giving up craving. He called it the noble eightfold path consisting of right view, right intention, right speech, right action, right livelihood, right effort, right mindfulness, right concentration. Thich Nhaht Hahn explains these concepts thoroughly in his book "The Heart of the Buddha's Teaching: Transforming Suffering into Peace, Joy, and Liberation."

Until the modern era beginning at the end of the 18th century with the discovery of a preventative cure for small pox, we Europeans lived with disease that was not materially different from what the Buddha experienced in his lifetime. Emotional and physical pain was caused in childhood by diseases like diphtheria, pertussis (whooping cough), and tetanus, sometimes fatal diseases usually found in children. About a million cases a year of diphtheria were believed to have occurred before 1980 world wide with a mortality of 5-10%. Small pox, measles, chickenpox and polio contributed to the experience of emotional and physical pain. These diseases have been almost completely eradicated in the modern era due to the discovery of vaccines beginning with Edward Jenner's discovery of smallpox vaccine in 1796. In

1928 Alexander Fleming discovered penicillin and it became possible to treat bacterial infections that caused physical and emotional pain.

What is the cause of emotional and physical pain in our modern world? The answer is: mortality. It has been said that life is a fatal disease, and there is no known cure for it. Awareness of our own mortality creates a deep anxiety that we keep largely out of our awareness. It is an extremely powerful force in western societies that have not resolved the issue. We will discuss death awareness at length here because understanding it is essential to understanding how civilization has ended up causing its own death and to understanding the need for Requiem.

About 50,000 years ago it appears, we humans became aware of our own mortality. We don't know if whales, dolphins, elephants, and the great apes are also aware of their own mortality. Awareness is an emergent function, and apparently it took 200,000 years from the appearance of the first Homo sapiens for humans to become aware of their mortality. My assumption is that it took all that time because it took that long for language to develop. As language developed, the brain was structuring itself. Eventually the sense of past and future emerged and then a human noticed that the fact that other humans died meant that one day he would die. That's when suffering began. In all the intervening years humans have suffered from disease, from war, from famine, from abuse, from murder, from loss of wives or children or parents. Always underneath the other sufferings, there was the ultimate suffering, the anguish from knowing one is mortal, that one will die one day. I believe we can date the appearance of mortality awareness to 50,000 years ago because that is when humans began to slaughter the Neanderthals and all the large carnivores as shown by the archeological record. I'll say more about this shortly.

In modern times, knowledge of our own mortality is the origin of the anguish the Buddha was referring to when he said all life is pain, the dictum he expostulated in the first noble truth. Humans have suffered from awareness of our own mortality. Awareness of death develops over the first decade of one's life. Some children learn about death from the nighttime prayer "If I should die before I wake I pray the Lord my soul to take." Some children learn when an aged relative, a grandparent perhaps, dies, and the child is exposed to the reality of the mortuary, the casket, and the dead body. Some children living in war zones or cities with gun violence or in families where murder has occurred see death first hand at an early age. The experience can be traumatizing especially if the child is not supported mentally and emotionally. Somewhere in adolescence young humans complete the process of dealing with their mortality, finding some way to go on with life while knowing it will end. Psychologists have studied the phenomenon of death awareness and humans' attempts to deal with it in a field called Terror Management Theory.

Awareness, especially awareness of one's own mortality, is too much for almost all human beings. Like the mole in the movie *El Topo*, we tunnel up out of the darkness of the unconscious and burst into the light and we are blinded. Because the light of self aware self consciousness is too great to bear, we blunt it. We blunt it with pure denial, unconsciously stopping any thought of death. Work provides a way to avoid self awareness which entails awareness of our mortality. We create corporations to serve so we do not have to serve ourselves. We drink and use drugs and watch TV. When people's accustomed defenses against self awareness are removed and they are placed in a situation where self awareness is more likely, they are often lost. They cannot imagine what to do. They become irritated. A vacation at a mountain lake where there is no cell phone reception and no TV can elicit the irritation that is a sign of discomfort. Just looking at the water, listening to the birds, raises an uncomfortable self awareness and with it an unconscious awareness of one's own mortality. Death anxiety is the fundamental cause of the Contagion that has brought the world to the brink of destruction. Humans are destroying themselves and the world we live in because we fear death.

The American psychologist Ernest Becker explored humans' relationship to death in his book, "The Denial of Death." Becker's book became the basis for Terror Management Theory which describes how humans deal with the awareness of their mortality.

Terror Management Theory

The human brain is capable of removing unpleasant, uncomfortable, or threatening topics from awareness through a process called dissociation. Thoughts of death and awareness of death are often dissociated out of awareness. Other names for dissociating an uncomfortable thought out of awareness are *repression* and *denial*. Pushing a threatening thought out of awareness protects the integrity of the psyche. However, dissociation has a down-side. Dissociation prevents a coordination between conscious thoughts and unconscious schemas. Mental flexibility is lost. Terror of death overwhelms the mind's ability to soothe the organism and prevents the organism from adaptively resolving the terror. Dissociation is a maladaptive mental response to a powerful threatening reality. Dissociation deals with an unpleasant thought by removing it from awareness. This book is spending so much time on the subject of awareness because dissociating death out of awareness is a central causative feature of the Contagion that is destroying our world.

In addition to just plain dissociating thoughts of death, humans have developed many other modalities to redirect their awareness of death. If the brain is occupied with other projects, it has no time to think about death. "Immortality project" is the term referring to a symbolic belief system that provides the self with a sense of superiority to physical reality. When the person surrenders to the immortality project, the person becomes a part of something eternal. The physical body may die, but the essential self, the soul, will not die because it has an eternal meaning conferred by the immortality project. The immortality project provides believers with meaning, purpose and significance and most importantly of all, it provides them with immortality and reduces the fear of death. The next few paragraphs will examine some examples of immortality projects that are important to understanding the Contagion. The history of our species, the succession of civilizations across the last 4,000 years is a record of our attempts to reduce the terror of death.

Religion has been the pre-eminent immortality project for most of the millennia of human history. The oldest literary work, written on tablets of clay in 2100 BC, tells the story of Gilgamesh who ruled the city of Uruk in ancient Mesopotamia from about 2800 B.C. Gilgamesh's story has all of the elements of an immortality project. In Gilgamesh's time people worshipped Anu and Enlil, among other great gods of heaven. Gilgamesh was cruel and arrogant and insisted on the right to have sexual intercourse with all new brides. In the epic, Gilgamesh has an upsetting dream that he will die one day. Enkidu is Gilgamesh's brother, a wild man, whom the gods kill when he and Gilgamesh slay the Bull of Heaven. Enkidu is a metaphor for Gilgamesh's unconscious. Gilgamesh searches for immortality and finally finds a wise man who advises the king to accept his mortality because he cannot change it.

People of ancient Mesopotamia found meaning through identifying with the story of Gilgamesh whose search for immortality mirrored their own. The *Epic of Gilgamesh,* which tells the story of the historical part-man part-god, provided ancient people with a symbolic belief system that enabled them to transcend ordinary reality with its knowledge of death and its accompanying anxiety. The story of the king gave meaning and purpose to people's lives, and through living out the story which was much larger than day-to-day life people obtained a sense of immortality. Notice that in the time of

Gilgamesh's story, humans had not yet invented the idea of a life after death. The wise man advises Gilgamesh to accept his mortality, very reasonable advice, which if humans had been able to accept it might have saved millions and millions of lives.

About 2,500 years after Gilgamesh, human civilization created a new solution to the problem of mortality: life after death through communion with the Savior, Jesus Christ, the son of God who was sacrificed to take on the sins of mankind and to ascend to heaven to sit beside God for all eternity. This solution was embodied in the canon of the Catholic church and has allayed the death anxiety of tens of millions of people. Catholics merge with the canon, which gives their lives meaning and purpose and a sense of immortality. When Catholics accept the eucharist in the ritual of communion, they are merging with the body of Christ. The canon says that believers will have eternal life after the physical body dies. Religion, whether Catholicism or its younger version Islam, palliates the terror of mortality.

In addition to religion, humans have imbued reproduction with the qualities of an immortality project. Humans unconsciously act as if reproducing themselves will assure their immortality, that they will live on through their children. Fathers believe their names will live on through their male children, although increasingly in these days of hyphenated last names, both mothers and fathers believe their names and hence some aspect of themselves will survive death. Genetics supports these beliefs because it demonstrates how through chromosomes and DNA the parents genes are passed to the children. Consider what relatives say about children: "He has your eyes," "She looks just like her mother," "He's just as smart as you." People are driven by an unconscious belief that they will survive through their progeny. This unconscious belief among all societies has resulted in overpopulation of the planet. Because the planet is overpopulated, food is becoming scarce in some places and will become scarce everywhere soon. Overpopulation and scarcity of food are contributing to the world dying. Reproduction is a form of immortality project for humans.

The corporation is another human construction imbued with immortality. The early forms of the corporation were the English and Dutch trading companies of the early eighteenth century, but in 1790 America invented the corporation and it immediately succeeded. Corporations are immortal by legal fiat. They cannot die. They also have the quality of personhood. The corporation is a perfect immortality project. People identify with the corporation to the point of merging their identities with it. It was once common for people to say, "I've been a General Motors man for 35 years." When workers join a corporation they participate in the immortality mystique of the organization. Working for Ford or Bethlehem Steel or General Electric gives people a purpose. Membership organizes their lives. Corporations have no feelings, no morals, no compassion, no empathy. The corporation exists solely to maximize profits for ownership. Having no feelings or morals allows corporations to do monstrous acts without remorse, acts that harm workers, citizens, and the environment. Profit functions much like the eucharist of the church's communion. By sharing in the profits, workers merge with the body of the immortal organization. The corporation provides medical benefits and retirement thus assuaging workers' fear of death. The corporation has taken a place next to religion among the immortality projects humans have created.

The bureaucracy is the twin of the corporation. Bureaucracies do not die. Governments are bureaucracies. Governments persist beyond the lives of the founding fathers and the statesmen who populate them. The United States government is a gigantic bureaucracy. Statesmen, politicians, and government employees participate in the agency's immortality mystique. The government provides medical benefits and retirement, thus easing people's death anxiety.

Merging with the bureaucracy organizes the employees lives and gives them purpose and meaning. Often when they die, they receive a plaque memorializing their service. The founders of the American bureaucracy have received an exalted position because they are held in awareness for many years after their deaths. George Washington is celebrated as the father of the nation over 200 years after his death in 1799. American citizens are indoctrinated into the canon of the American immortality project, and some of them will go on to serve the project and sometimes die in service of the immortality project. Service to the nation and merging one's identity with the nation provides a sense of immortality because the nation is believed to outlive the citizens who serve it.

Reproduction comprises an immortality project, and the family is the basic social unit within which reproduction takes place, and it is therefore the fundamental immortality project. People identify with their family and say, "I am a Smith," or "We're the Thompsons from Missouri." After the family, humans' next higher level of organization was the clan, then the tribe. We humans evolved in clans of a few dozens of people. The clans assembled into tribes of a few hundreds. Clans and tribes constituted our basic organizational structure, and they also functioned as immortality projects. As a species we succeeded in part because we owed allegiance to clans and tribes. Clans and tribes gave meaning to our lives in ancient times. The clan and tribe helped alleviate the death anxiety. Early humans hunted in packs and defended themselves in clans and tribes. Clans and tribes provided protection against starvation and being killed in war. As civilization evolved, people shifted their allegiance to their village, their shire, their country, their nation. In modern times, people have shifted their organizational allegiance to corporations and to bureaucracies. The corporation has subsumed the family, clan, and family. Corporatists believe their survival depends upon survival of the corporation. In a contest for survival of the fittest, the corporation is more fit than individual humans, families, clans, or tribes. Corporations have supplanted clans and tribes as the dominant immortality project in modern times.

Wealth can become an immortality project. Accumulating wealth appears to provide a defense against death anxiety. Poor people live with fear that a medical bill, a car accident, a speeding ticket, an insurance payment will push them into poverty or homelessness where they are more susceptible to hunger, weakness, disease, and death. Wealth accumulation confers a protection against these dangers which are reminders of mortality. Accumulating great wealth, multi-millions or even billions, is more than a defense against the vicissitudes of life. Great wealth is an immortality project in and of itself. Great wealth gives meaning and purpose to the wealthy person. Wealth confers specialness. The wealthy person is believed to be favored by God or fate. It establishes the wealthy person in the pantheon of the truly successful, the cultural maximizers of the age or of all ages. The wealthy person knows that when they die, their memory will live on in the form of charitable gifts to institutions, inheritances given to relatives, research institutions bearing their name, buildings named after them. The wealthy person of today identifies across the centuries with the wealthy Italians of the Renaissance, or of the French monarchs of Versailles, or of the patricians of ancient Rome. Wealth accumulation is a significant immortality project.

Cultures and societies have erected buildings as defenses against death anxiety because the financiers, architects , and laborers all know that the edifice will survive them. Designing and building and paying for the structure gives purpose and meaning to the lives of all involved, and everyone who observes the structure participates in the mystique of the immortality project. What attracts people to the great pyramids of Egypt if not the sense of participating in an effort, a religious project, carried out long ago that survives to this day? Skyscrapers, temples, cathedrals, ziggurats, and pyramids in meso-America as well as Egypt comprise immortality projects. Great museums, which sometimes have a patron's name associated with them like the Guggenheim in New York, the de Young in San Francisco, or the Norton Simon in Los Angeles, also constitute immortality projects. The great pyramids were associated with the pharaohs entombed there, although the names of the pharaohs were lost to our collective awareness and have only recently been raised to awareness through the efforts of archaeologists. For the most part though, the names of individuals who planned and labored on these structural immortality projects are unknown, subsumed by the grandeur of the building itself.

Not all immortality projects are as grand as the pyramids or the Eiffel Tower. Some immortality projects are personal and paradoxical. Drugs and alcohol provide a means of chemically dissociating the anguish of knowing one will die, and hence they qualify as an immortality project. Alcohol induces a blackout where emotions are numbed. Opiates dull the fear of death, replacing it with a chemically induced merger with the "All," a perfect attachment to the great mothering universe. All substances of abuse provide a paradoxical solution, because while they provide a temporary relief from mortality awareness through intoxication and blackout, the intoxication and blackout are enactments of death. Blackout is a state of non-being. Opiate intoxication is a state of non-being. The high of opiate addiction or of alcohol-induced blackout is always followed by an extremely painful withdrawal. "I feel like I want to die," wails the blackout alcoholic whose head is pounding. The heroin addict cycles between shooting up and coming down, unconsciously reenacting the cycle of death awareness and brief respite from death awareness. The anguish of knowing one will die is so painful that addicts and alcoholics seek emotional relief in substances that shorten their lifespan. Death anxiety is buried so deeply in the unconscious that addicts and alcoholics rarely become aware of its motivating effect on their addiction.

During Roman times society developed yet another means for allaying death anxiety through the vicarious deaths of others, the gladiators in the circus. Spectators gained power over death because they were ordering the deaths of the combatants. Spectators participated in the death mystique without actually having to risk death themselves. Vicarious experiencing of death has persisted across the ages in many cultural rituals. In the American west of California, bear and bull fights were staged in which a grizzly bear fought a bull to the death. These events, held on Sunday after church services, usually resulted in the death of the bull. The bull had only one chance to win, which was to charge across the small fenced arena and gore the grizzly with his horns. If he failed to score, the bear would leap on the bull's back and bite his neck. Crowds cheered. Children were exposed to the spectacle. These vicarious enactments were often preceded by cockfights or dog fights which although illegal persist in rural areas to this day. Public hanging and beheading and draw-and-quartering consecrate death in the realm of law, courts, and the power of the state to take life. Citizens watching these events brought death under control. The lesson learned was to obey the law and avoid death. The deeper meaning was immortality: The Roman circus, the bull and bear fights, and public executions all confer power over death, thus allaying society's death anxiety.

The contemporary era provides us with new forms for vicariously experiencing death: TV, films, and video games. Deaths on TV comprise a major means by which we humans moderate our death anxiety. Death is ubiquitous on TV. The average American child will witness 16,000 murders on TV by the time he or she graduates from high school. Witnessing a shooting, stabbing, throat-cutting, immolation, choking, or beating to death constitutes a vicarious experiencing of death that diminishes the terror associated with our own death.

Viewers know the death is artificial. Viewers participate in
the mystique of the death, and they have power over death
because it can be staged. When death happened before
contemporary times, it was not artificial. It was actual.

American society in particular has developed another
means for defending against death awareness: accumulating
weapons as individuals and as a society. The numbers of
weapons designed for killing that are owned by Americans
reflects American society's deep fear of death. Smith &
Wesson is one of the pre-eminent American arms
manufacturers. The company assessed the domestic market
for auto-loading long rifles—no one likes to admit these are
assault rifles because mass murderers like Adam Lanza who
killed 26 children and teachers at Sandy Hook in 2012 have
given the weapon a bad name—at $489 million dollars.
Researchers estimate there are 310 million firearms in the
U.S. including 114 million handguns, 110 million rifles, and 86
million shotguns. Estimating the number of assault rifles has
proved more difficult. Analysis suggests that Americans
possess 3.25 million foreign and domestic assault rifles.
Manufacturers don't like calling these assault rifles because
they are supposedly sold to kill someone attempting to kill the
home owner, although they are more often used in mass
murders. The manufacturers would like Americans to think of
their product as a protective rifle or shield rifle. The facts belie
this claim. Burglary is the main cause of someone entering a
home. Data show that 61% of burglars are unarmed. Burglars
are more likely to be known to the victim. In the period
between 2003 and 2007 there were no deaths due to home
invasions. Yet citizens accumulate handguns, rifles, and
shotguns. The enormous number of weapons of all kinds—
one weapon at least for every citizen in the nation—reveals
American's mortality anxiety, the belief that at any moment
armed intruders could break down their doors and kill the
citizens and their families. Underneath this belief lies an
unconscious anguish about death and dying. No thought, no

fact can change the unconscious fear of death. It must be raised to conscious awareness and then and only then can it be resolved.

The paradox of keeping weapons to protect against home invaders is that the weapons are far more likely to be used to harm the home owner or someone living in the home or a visitor. On average 13 people age 10 to 24 die from firearms every day in America. None of these young people was invading a home. Children shoot each other accidentally when playing with a loaded firearm while parents are out of the house. Firearms are much more likely to be used in suicides or murder of a family member. Romantic partners use handguns to kill their former partners. Between 2003 and 2014 10,018 women were murdered in 18 states. Fifty-four percent were gun deaths and 55% of the murders were committed by a former or current partner. Children take unsecured handguns to school and shoot other students whether accidentally or on purpose.

As a globe-spanning civilization, humans collectively invest their wealth and the fruits of their labor in accumulating vast amounts of military weapons. America's fear of annihilation is so great that it spends more on defense than the next seven countries combined. A deep dread of death motivates the nation to invest in weapons. The nation spends its wealth on imagined dangers rather than on very real dangers like crumbing infrastructure, inadequate health care, education and schools, or mitigating the effects of global climate change. The money spent on weapons enables society to commit its most deadly defense against death anxiety: externalizing fear of death through murdering others.

Externalization is a psychological defense against anxiety. A person who fears death externalizes that fear by getting someone else to feel it. The fear of death is projected onto the outside world. For Americans, the externalizing defense is excessive and has led to the development of a cultural neurosis. Americans murder each other, and America murders human beings in other countries. In that way, America induces others to feel the death anxiety that we hold deep in our unconscious. Awareness of death is traumatic if unresolved, and the trauma begins in childhood when knowledge of death first occurs. Psychologists have long known that we humans reenact the traumas we've experienced in childhood, often in disguised form. Humans reenact the trauma of awareness of death in the disguised form of murder, of killing. In the act of murder or killing, the killer induces the victim to feel the dread of death, and then the murderer does not have to feel that awful anguish. Murder appears to convey power over death. Murder externalizes the murderer's death anxiety.

The state is often the agency of killing in the case of execution, in the case of warfare and in the case of police killings. Between 2014 and September, 2018 America executed 121 citizens. Texas executed 45, Georgia 19, Missouri 18, and Florida 15. For comparison during all of 2017 1,940 humans were executed world-wide with China the pre-eminent executioner, killing 1,000 human beings. American executions are ritualized. At trial a jury finds the defendant guilty and then recommends a sentence. The judge imposes the sentence of death. There follows an appeal, another decision, and if the prisoner does not prevail they are placed on death row, a special group of cells that separates the condemned from the rest of the population. Currently the nation has been discussing how the prisoner will be executed, whether by lethal injection, electrocution, gas, hanging, or firing squad. The discussion has focused on minimizing the dying person's suffering. As the day of the execution approaches the news will report the prisoner's last meal, the visit by a priest or minister, the prisoner's last words. Witnesses are seated in a viewing area separated by glass from the death chamber. The prisoner walks the last few yards from a holding cell to the death chamber escorted by armed guards. The sentence is carried out, and a physician certifies that the prisoner is dead. Witnesses often describe the prisoner's last moments for the press. Usually only first degree murder qualifies for the death penalty. The condemned person has imposed death on another, and a murder like that raises the death anxiety of the populace. First degree murder is premeditated murder which means the killer planned the crime, but the victim was probably unaware the crime was forthcoming. Citizens unconsciously identify with the victim and worry their own death could be random, unforeseen, and imminent. The capture of the murderer, the trial, the sentence, and the ritual of the state-sanctioned execution of the criminal give citizens a sense of power over death. The entire sequence creates meaning for citizens who believe that justice will prevail and murder will be avenged by the State. State-sanctioned execution externalizes citizens' death anxiety. Citizens imagine what the prisoner must be thinking and feeling knowing that he or she is about to die. Most of us do not know that we are about to die. Even terminally ill patients know they are "going to die," but they do not know that in five minutes their awareness will cease. State-sanctioned executions are a means for easing society's death anxiety.

Nations engage in warfare that externalizes death anxiety through killing other human beings. Externalizing death is unconsciously motivated, so citizens are unaware of the unconscious schemas that prompt their behavior, but the causal relationship exists whether aware of the connection or not. Psychologists have learned to decode the contents of the unconscious by examining the current behavior and inferring the contents of the unconscious. The sheer volume of murdering that occurs in the world attests to the murderous power of the unconscious. The conscious mind will always create an explanation for the unconsciously motivated behavior, because the conscious mind cannot accept the reality of the behavior. The following paragraph lists some examples of humans murdering other humans. Chapter Three will explore these in more detail.

Human beings murder each other. Whites murder blacks, blacks murder blacks, and whites murder whites. White police murder blacks. American men murder their wives and sometimes their children. In Mexico gangs or cartels like the Sinaloa Cartel or the Tijuana Cartel murder each other and citizens. In Africa Hutus murder Tutsis, and the Lord's Resistance Army murders villagers and the unprotected. In Myanmar, Muslim Rohingya are murdered by Buddhist Rakhines. At one time Catholics murdered Protestants. Irish murder each other. Israelis murder Palestinians. Nazis murdered Jews. In the Middle East ISIS murders Muslims who do not support ISIS. Women are raped and murdered in India and sometimes Indian parents murder their daughters for having sex outside of marriage or for marrying a male the parents don't approve of. In China Tibetans are murdered. In America and in many places in the world Lesbians and gays are murdered ostensibly because of their sexual preference.

In each of these cases the existence of the Other— whether black, white, Jewish, Muslim, Sinaloan, Palestinian, Hutu, gay—is a threat to the identity of the murderer. The threat to the murderer's identity feels like death. The victim's existence reminds the murderer that he is impermanent, that his identity could die, and so the murderer externalizes his fear of death through the act of murder. Killing someone perceived as Other, gives meaning to the murderer's life. In the case of "honor killings" in India, the murdering parents or the brother who has murdered his sister's lover believe they are part of a system that is larger than the self, a system of honor.

In addition to murdering each other, humans murder the environment. Humans experience death anxiety that the environment could kill us. Typhoons, tsunamis, hurricanes, tornadoes, avalanches, floods, and earthquakes are some of the ways that the environment kills humans, and humans fear death could come without warning from one of these causes at any moment. Human beings are unaware that their behavior towards the environment is murderous. Unconsciously motivated murderous behavior towards the environment is paradoxical because we need a healthy environment to live, and yet we murder the oceans with plastics and chemicals and radioactive waste and agricultural run-off of herbicides and pesticides and fertilizers. "Dead zones" where oxygen has been depleted and no animals can live and plants die off result from murdering the oceans. Humans murder our supply of clean water by injecting toxic waste from fracking into wells. We murder the air we breathe with toxic smoke from factories and toxic exhaust from automobiles, trucks, buses, and airplanes. We murder the land with garbage in landfills, with radioactivity as at Chernobyl and Fukushima and Three-mile Island, with herbicides and pesticides. We also murder the wild animals who live in the environment. Coyotes, wolves, bears, elephants, rhinoceros, giraffes, ducks, geese. The environment and the animals who live there are wild and free. They are part of a natural system of birth, life, and death, and that natural system threatens humans because we cannot join with the system without accepting our mortality. Humans are unable to find meaning for our lives by joining with the environment and the animals who live there, and so we discharge our anguish through the mercilessly angry act of murder.

Before moving on to Chapter Three, a brief summary of the book's teaching so far is in order. Chapter One described awareness and its importance to understanding what is happening to the earth, and to the need for Requiem. Chapter Two developed essential thoughts about awareness of death. Take a moment and reflect on what thoughts, images, memories, and emotions came up for you as you read this chapter. Did you feel uncomfortable with such an extended discussion of death? Was the idea uncomfortable that we humans externalize our death anxiety in many ways including murder? If you were uncomfortable you are experiencing what it feels like to pull an unconscious schema up out of the depths of your mind and to bring it to awareness. In Chapter Three we will investigate the attitudes and behaviors that have arisen from our unconscious fear of death.

Chapter Three

The Contagion

What is killing us?

Human anguish at the prospect of our own death has given rise to a complex of behaviors that act as defenses against mortality awareness. The group of behaviors is united by fear of death. These behaviors and attitudes and things include: money, property, greed, dominance, violence, entitlement, and self-absorption. Collectively they are called Contagion. In the brain they form a schema, a mental codification that organizes perception, thought, emotion, and behavior. The schema is unconscious, meaning that people are largely unaware they are infected with the schema.

Each of the Contagion behaviors and attitudes is the negative pole of a pair, and each has a positive pole, an adaptive attitude or behavior. Greed is the negative pole of the dyad in which gratitude and sufficiency form the positive pole. Greed is a defense against the emotion of disgust. When humans become better aware of their emotions they recognize they feel disgust—the emotion of rejection, the emotion that says something is a contaminant—at what our species is doing to the world. Money externalizes competition and it prevents the development of mutuality, or working together for a common cause. Property is the idea that one can own or possess a piece of the commons that is the earth and that could sustain us all. The idea of property is the negative pole of the commons, the unity of all the surface of the earth, none of which belongs to anyone. Property Is a means for externalizing competition, and it is the negative pole of the pair that includes cooperation on the positive side. Self-absorption is the negative pole of compassion. Dominance at the negative end prevents collaboration at the positive end, and violence prevents the positive behavior of peaceful resolution. Collectively the seven elements of the

Contagion manifest humanity's defenses against death anxiety. The Contagion is also the cause of the death of our world.

How the Contagion is perpetuated

The Contagion manifests in the form of perceptions, emotional responding, attitudes, thoughts, and motivations for action that are stored in the unconscious and are collectively called schemas, another name for plans. Understanding how the Contagion is passed from generation to generation requires knowing something about the human unconscious. During the first four or five years of life infants are absorbing vast amounts of information about parents, siblings, people, relationships, how to stay safe, how to get their needs met. Every observation is stored and becomes the basis for subsequent behavior.

Infants' brains have not yet formed sufficiently to be able to create stories or narratives about what they are learning. The stored information becomes the unconscious because we don't have memories in these first years of our lives about ourselves and how we learned what we learned. The information absorbed during this time affects our perception, our thinking, our feeling, and our behaving. The only way we can know what we learned in these early years is by analyzing our current perceptions, thoughts, emotions, and behavior. Much of what infants learn they learn from watching their parents. If father is loving and kind to mother, the infant learns that adult males are loving and kind to their partners, and later in life he will treat his own partners with love and respect and kindness and compassion. Their partner may say, "You are such a kind person. Where did that come from?" and the person will say, "I don't know. I guess I've always been like that."

By the same token, if father was angry and verbally and perhaps physically abusive to mother, the infant learns that men treat their partners with disrespect, meanness, jealousy, and verbal and perhaps physical abuse. Later in life that man will be more likely to lash out verbally and perhaps physically when he feels frustrated, when he doesn't get his way. He will act entitled, just as his father did. His partner may ask, "Where does this come from? One minute you're nice and the next you're calling me names." The man will say, "Shut up. I don't know. Something comes over me." In neither of these cases does the man know what motivates him to act as he does. His unconscious is directing him. The only way the angry man will ever change his behavior is to become aware of his angry behavior, to raise it to awareness. Unfortunately if he were to become aware of his behavior he would have to be able to deal with the shame that arises when he does become aware of his behavior. Shame is a powerful opposing force to awareness and to behavior change.

The unconscious schemas are carried in two brain pathways that are very fast, automatic, and not readily reviewable by the parts of the brain that can assess how well a behavior is working. A schema-driven behavior is difficult to recognize. Over evolutionary eons, the unconscious pathways have evolved with one objective: to ensure the survival of the organism. There is only one conscious pathway in the brain that can assess how well the unconscious, schema-driven pathways are working, but that pathway is ponderously slow in comparison to the schema-driven pathways. By the time the self-reflective pathway is engaged, the schema-driven behavior has already occurred, and defense mechanisms of denial, justification, and verbal and physical violence are starting up to protect the self against the painful feelings of shame and guilt. To reiterate, the infant constantly learns by observation of its parents and others and the world. The infant has one motive: to survive. What the infant observes is stored in the unconscious where it becomes the basis for future thought and action. The infant's fear of death becomes the adult's death anxiety.

For humans to change their current behavior, as individuals or as a society, they would have to raise the unconscious schemas to awareness. Psychotherapy is a process whereby unconscious schemas are raised to awareness. Once raised to awareness the self-reflective circuit can assess how well the schema-driven process is working. Change is then possible. The person has to become aware in order to develop new responses. During all of this the defense mechanisms are activated. Change feels like death because an established part of the personality is going to have to die in order to be replaced by a new and healthier part. Denial is an unconscious schema that opposes raising unconscious schemas to awareness.

Neuropsychology can now describe the operation of unconscious, rapid, automatic, emotion-driven schemas. These schemas motivate racist behaviors, misogynistic behaviors. They motivate terrorism, fascism, spousal abuse, child abuse, business enterprise, polluting behavior, the tragedy of the commons, war, and global warming. Neuropsychologists understand why people who "know better" keep re-enacting the same old patterns. The "knowing better" is a frontal cortex function. It is slow, ponderously slow compared to the speed of the unconscious schemas. The frontal cortex assesses what works and what does not. From the point of view of the unconscious, the Contagion behaviors are working just fine. The reward is immediate. Dopamine goes up. Utilizing the knowledge of the frontal cortex for change entails work and effort. There is no immediate reward. The dopamine coefficient is low compared to the dopamine reward of the unconscious schemas. The unconscious schemas do not consider the future. Knowledge of the damage the Contagion is causing is immanent. It has not yet been raised to awareness in a sufficiently large portion of the human population to result in change. Tragically awareness will not emerge in time to save the world and our species. This is why we must now perform a Requiem.

Psychologists have studied the transmission of many adverse human behaviors from one generation to the next. Spousal abuse is often intergenerationally transmitted. Boys learn from their parents to be spousal abusers when they grow up. Boys learn that men dominate and women submit and resistance causes violence, and they learn that men deny their behavior at the same time as they feel entitled to that behavior. Girls learn that relationship entails being emotionally, mentally, physically, and sexually dominated. Girls learn to submit or be harmed, and they carry this knowing into their adult relationships.

Greed, pursuit of money, dominance, unlimited expansion, racism, bigotry, violence, attitudes toward the environment, all of the attitudes of the Contagion are all based on unconscious schemas learned in childhood. The attitudes that make up the Contagion are passed from generation to generation. Psychologists call this "intergenerational transmission." When enough people are infected with Contagion a population effect develops, and the attitudes of Contagion become attitudes of the society. The American psychologist Murray Bowen described a phenomenon of "cultural momentum" to help us understand how attitudes are transmitted from generation to generation with such force.

Gun ownership in America provides an example of cultural momentum and intergenerational transmission. Guns are given to boys as toys, but rarely to girls. Data are difficult to obtain, but it appears that fathers transmit attitudes towards guns to their sons. Boys form an unconscious schema about gun ownership that is resistant to change. For evidence consider the attitudes of the N.R.A. and gun owners towards their weapons. For boys and men who do own guns their masculinity and personal power appear merged with the firearms. Data on handgun ownership in two different countries illustrate the power of intergenerational transmission and cultural momentum. In the U.S. every 100 people own 88.6 guns. In the U.K. there are 6.2 guns per 100 people. The frequency of gun ownership in Japan is 0.6 per 100 people. Different populations can have different schemas regarding gun ownership resulting in a different cultural momentum in each society.

To better understand the concept of cultural momentum, imagine that the Contagion is a gigantic freight train where each freight car is a Contagion-infected individual. The Contagion-infected train has so much mass—so many cars—and it is going so fast that it has gathered enormous momentum. The Contagion-infected train has gathered so much momentum that it blasts aside protesters, arguments, awareness of the effects of the Contagion like drought and global warming and pollution and melting polar ice caps. It runs right over science and observations and reasonable objections. Changing the momentum of the Contagion would require a slow, incremental process, something like psychological brakes reducing the speed of each car. Unfortunately there is no opposing force, no brakes to slow each car. The momentum of Contagion has run over and obliterated every opposing force. Looking at voting records to assess the momentum it appears that about 30 percent of Americans are deeply infected with Contagion. There is no opposing force held by enough citizens to stop the momentum the Contagion has built up.

When societal change does occur it is a multi-generational process. The parents of today recognize one of their unconscious schemas is not working, they raise it to awareness, change the behavior, and then they will model new, slightly better lessons for their children who will in turn model and teach somewhat more adaptive lessons to their children. American society resists change and when it does change at a surface level the deeply rooted attitudes often do not change to match the surface change. An example is American society's attitude towards women. At the founding of the United States, women were so subservient, so marginalized that they are not even mentioned in the

Constitution, and they were not granted the right to vote until 1920 when the Nineteenth Amendment was passed. Women were treated like chattel. Still today women are having to fight for equal pay for equal work. The problem for our dying world is that we do not have the luxury of dozens of generations to heal the Contagion.

People in recovery from addiction deal with the challenge involved in needing to make an immediate change in an established behavior in order to survive. Recovery asks them if they've become sick and tired of being sick and tired. Step One of the Twelve Steps advises admitting one is powerless over the addiction and that life has become unmanageable. Often recovery requires a moment of clarity when the consequences of continued drug or alcohol use are so apparent that the person becomes willing and open-minded and honest enough to begin to change the unconscious schemas that have driven their addiction. The devastation of hurricanes Florence, Sandy, or Michael does not appear to have broken through society's denial. We have not heard Senator Inhofe admit he has been wrong to deny climate change. Can our species become so unhappy, can our lives become so unmanageable due to drought, rising oceans, scarcity of food, and temperatures rising more than 2 degrees Celsius that humans will seek change?

This chapter presents the information necessary to make a decision to change. People in recovery say "Sometimes the addict has to die" meaning that for some addicts the unconscious schemas motivating their addiction are stronger than any conscious, self-reflective motivation. Homo sapiens could be a dying addict. That is the reason for Requiem. The idea that all human beings could raise their schemas to awareness and overwrite them seems inconceivable. The following sections review the ways the Contagion has manifested in our world. These comprise the unmanageability of our world, the consequences of infection with Contagion. The unconscious schemas driving these behaviors must be changed by most human beings if we are to survive as a species.

Contagion and Treatment of Indigenous People

The Contagion has many names: American exceptionalism, white privilege, male privilege, anthropocentrism. The Contagion destroys Theory of Mind, the human capacity to be aware of what another is thinking or feeling. Europeans infected by the Contagion could not feel compassion for the suffering they were wreaking on the Indigenous people. Europeans destroyed the culture of Indigenous peoples, taking their land, laying waste to their villages, forbidding the spiritual practices of Lakota, Sioux, Comanche, Choctaw, and every other tribe. Children were taken from families and placed in indoctrination camps called "reservation schools" where they were forbidden from speaking their native tongue and were physically, emotionally, and sometimes sexually abused. Europeans would feel anguish if their own children were taken from them, but they could feel nothing for the anguish the Indigenous people were feeling. Indigenous people experience a oneness with their homelands. European invaders could not feel compassion for the emotional pain the loss of their homeland imposed on indigenous people, because the Contagion had destroyed their Theory of Mind.

Contagion and Slavery

The Contagion also prevented European slavers and slave owners from experiencing the emotional pain of the African people they took from their tribal culture, turned into slaves, and then beat, raped, lynched, and sometimes immolated. Posters advertising sales of slaves in 1855 referred to the men as "bucks" and to the women as "wenches." "Buck" and "wench" are Contagion names that prevent awareness that the people being sold were human beings who had been traumatized by their capture and removal from their villages, families, clans, and tribes.

Contagion and Deception

Contagion taught the Europeans to practice deception. Contagion manifests as deception, cheating, lying, and manipulation in the perpetrators. The Indigenous victims of European deceptions were unable to recognize those deceptions of the perpetrators. The Indigenous people were naive when Europeans arrived and offered them treaties. Because the Indigenous people had no experience with the Contagion they could not recognize the Europeans' lies and manipulations. Indigenous people were at a disadvantage because they were not familiar with deception. Europeans signed treaties with Indigenous people that were to last "as long as rivers flowed." Very quickly the Europeans broke the treaties. One deceptive tactic the Europeans used was to negotiate a treaty between the national government and a native tribe and then to create a state—this happened in Alabama—that controlled the land. The state then gave the land to speculators and settlers, thus cheating the native people. The federal government claimed it could do nothing because Alabama was a state and Washington had no power over it to make it honor the terms of the treaty.

Contagion-driven deception lives on in today's world. The Contagion has damaged right hemispherical cortex in the brains of both victims and perpetrators. The perpetrators feel entitled and greedy; the victims feel powerless and overwhelmed. The victims are unable to recognize the intentions of the manipulators, the conservative politicians who intend to harm citizens by taking away their Medicare, taking away their Social Security, taking away access to reproductive services, and by increasing taxes on the victims and reducing taxes on the 1 percent who are the perpetrators. Politicians have refined the practice. Most recently the government promoted its Tax Cuts and Jobs Act, promising the new law would benefit America's middle class. In fact the law cut the taxes of the wealthiest one percent of Americans. Eighty-three percent of the benefits flow to the wealthiest one percent. Middle class Americans will actually pay more taxes within a decade under the new law. Deception flourishes currently under the Contagion.

As we are learning, the Contagion prevents the attribution of mental states to others to be able to predict their behavior. Trump supporters do not attribute deception to Trump. Trump's spokespersons immediately appear on TV after Trump has revealed his true intentions and deny that Trump has revealed his true intentions. Currently, based on recent elections, it appears that about 30 percent of voting age Americans believe the lies of the politicians. Citizens infected with the Contagion are unable to recognize the deceit that is being perpetrated on them. Citizens who are unaware of themselves cannot recognize Trump's intention to enrich himself and his class of wealthy people.

Recognizing deceit is more difficult when the deceiver employs an authority trance. The authority trance is a method that uses the perceived authority of the hypnotist for inducing a hypnotic state. An authority is someone with a high level of knowledge or power, an expert. Authorities use commanding language to enhance their position. A hypnotist projects an air of confidence. The authority trance is often used to coerce or persuade a person to do something they would not normally do. Politicians are often perceived as authorities, and some of them use that power to induce a dazed state of absorption in which one's self-awareness is diminished. Awareness of deception departs along with self-awareness. Politicians have authoritatively stated they will protect the environment only to subsequently vote for a pipeline across ecologically sensitive land or for fracking after promising to promote alternative energy sources. Citizens who are being deceived will tell themselves, "Well, he's a United States Senator so he must know what he's doing." Citizens will then often refuse to acknowledge that they've been duped because to do so would cause them to feel shame. The citizens who voted for Trump for example and who continue to support him even though his lies have been exposed are people whose self-image is fragile and they would feel like they were dying if they admitted they'd been had.

Contagion and racism and bigotry

A battle is raging in the collective awareness of us human beings, a battle between the ability to model the experiences of other human beings beside the self on the one hand and the Contagion of money-property-greed-self absorption-dominance-violence on the other hand. The Contagion lives on in the attitudes of contemporary bigots and racists who have no compassion for the suffering of African Americans, Hispanics, women, and LGBTQ youth and adults whom the bigots scream at, insult, assault, terrorize, and control. The white haters would react with fury if their own children or relatives were treated the way they treat black, Hispanic or LGBTQ people. The Contagion prevents the haters from mentally modeling the anguish of the targets of their hate, because if they were to feel compassion they would also have to feel disgust and shame toward themselves for their behavior. Shame and disgust are too overwhelming to feel and the Contagion attitudes provide a justification for their hatred and a defense against shame and disgust. The following exercise will help you model the experience of the victim of racism. The exercise assumes you, the reader, are white.

Sit for a moment in the safety of the ritual space of this sentence and allow an awareness of the suffering of the target of racism to arise. Allow yourself to be the black man who has had the cops called on him for entering his own apartment building. Be the black girl who has set up a lemonade stand and has had the cops called on her. Be aware of your knowing of your own worth, and value, and ability which is being denied to you. Be aware of the emotions and sensations that come up when you are being marginalized,

*excluded, denied, and called names and sneered at. Be
aware of your knowing that this would not be happening if you
were white. Now, reader, notice your own defenses, your
feelings of entitlement. This will not be easy. You may have a
thought like this: "I am white and the society serves me
because America is a white nation." Notice your fear at any
thought of embracing the defensive attitudes and behaviors of
the black American. Notice, reader, your Contagion-
dominated fear of being attacked or losing your favored place
in society, of losing control. Give yourself plenty of time to
experience the discomfort, to notice the thoughts and
emotions that come up. Now, take a deep breath. Let it out,
and return to the place where you started.*

To whatever degree you were able to carry out this
exercise, you will have experienced both sides of Contagion-
driven racism, the victim's experience and the racist's
experience. Sit with this experience. Digest it. Embrace it.
Realize you have encountered the roots of racism: the threat
of death of the identity that racism and bigotry defend against.
Realize as well the enormity of the change that society must
undergo if we are to heal the Contagion and create a just,
healthy society, one that can repair all the damage to race
relations, to the environment, and to the world.

Contagion and Women

"Misogyny" is the name for hatred of women that Contagion
produces in men. For eons men have had power over women,
but this was not always so. At one time society was
matriarchal. Women owned property and held the positions of
power in society. The reason women held power was because
they could reproduce and men could not. Because women
could reproduce, they had power over death which
exacerbated men's death awareness. Reproduction was a
mystery that inspired fear in men. Sometime during the
Neolithic period beginning at 10,000 BC and ending between
4,500 and 2,000 BC, it appears men figured out the rudiment

of reproduction that men's semen was required for women to reproduce. With that men gained a portion of the power to create life. Men seized that power and began to dominate women. Men changed from revering women to perceiving them as dirty, slaves to be used at the whim of the man, servants, whores. Men began to fear women because they could take a man's power through their ability to incite lust in men and their ability to bring a man to orgasm. Men began to hate women. Some men beat women. Men denied women equality and the right to vote. In contemporary times, men sometimes replace women with younger and more desirable women. Men exert their dominance through rape. Men may even murder women.

Men's hatred of women arises in part from men's neediness for women. All children—but this is more relevant to men—are born in a state of neediness. Men need mother to nurse them, to feed them, to care for them, and to protect them. As male babies are socialized into a male identity they are taught to be independent, and as a result they come to hate the neediness they felt for mother. Neediness diminishes the sense of independence that holds the male ego together and provides a defense against death awareness. Not all men hate women, but contemporary societies are characterized by misogyny. Male babies want to control mother, to make her be the Good Mother who always satisfies the male baby's needs. They feel anger and anguish when they are unable to control mother. These emotions and impulses condition the unconscious in men and motivate their behavior later in life. Female babies often do not experience this kind of hatred because they are cis-gender with their mothers. If their mothers are controlling or assaultive the female baby is more likely to grow up hating women.

In the modern era women are demanding equality and men feel threatened. The integrity of men's ego structure is challenged by powerful women. Although few men would recognize it, when their ego structure is challenged men's mortality is threatened. They feel powerless, which feels like death. So men use misogyny to defend against their powerlessness. Men infected by the Contagion lack the ability to see themselves as they are. The shame of accepting the ways in which they have treated women would be too much to bear. Men infected by the Contagion try to regain power by controlling women's reproductive lives through closing Planned Parenthood clinics, denying abortion, denying healthcare, and abandoning the pregnant woman. Men act out their misogyny by forcing women to have unwanted children which further enslaves them. They deny the woman freedom over her own body. Men do it because they can. They punish women for being women.

The Contagion affects the thoughts, perceptions, attitudes, and actions of the fathers who recklessly impregnate a woman and then leave her. The behavior is called Baby's Momma in some places in society. It is a violent act of domination and control. The impregnating father takes advantage of a relationship that should be loving and caring. The male has a plan all along to leave, but he hides his plan, an act of deception. The woman is naive and believes the man's words are truthful. The man's actions have all the characteristics of Contagion. He is narcissistic and self-absorbed, caring only for himself. He is greedy for the sex and the power and domination. He cares little for the woman or the baby. The man impregnates and leaves. He has dumped his contaminating semen into a vessel, and leaves without a care for the well-being of the child he has fathered or for the anguish of the child's mother, the baby's momma. He deprives her of her life, 18 years of her life. He leaves her with the anguish and terror at having to raise a child as a single mother. He leaves society with the bill, expecting it to be paid somehow, but uncaring of exactly how. How like the actions of the businessman, the owner of the coal mine, the slave owner, the brothel owner.

Contagion prevents awareness of the suffering of women who are denied health care. Politicians who defund Planned Parenthood have no awareness of the suffering of a young woman who has gotten pregnant and has been abandoned by the father of her child. She may have been raped. The man may have deceived her, motivated by his own Contagion-driven male entitlement, telling her he loved her and would take care of her forever. As we learned in the previous chapter, Congress and politicians and society externalize their own fear of loss of identity and loss of control of their lives or their son's and daughter's lives through an unwanted pregnancy. The father's identity depends on having an ideal family, one that is free from what he perceives as the taint of unplanned pregnancy. Loss of that identity feels like a death. That father is more likely to vote for candidates who promise to close Planned Parenthood clinics. Furthermore, because he fears dying for lack of medical care, he will externalize his death anxiety by denying medical benefits to the poor or other citizens who have not succeeded well enough to provide for their own health insurance.

The Contagion affects thought processes, providing justifications, explanations, and excuses for the behavior of denying medical benefits or food to the impoverished single mother or abortion services for an unplanned pregnancy. Sit for a moment in the safety of the ritual space of this sentence, page, paragraph, and chapter, and experience for a moment your own thoughts that come up in this brief exercise.

Sit quietly. Empty your mind for a moment of thought. Now, allow an awareness of the suffering of the young single pregnant woman to arise. Feel the heartache as she realizes the man she loved and believed loved her has betrayed her, abandoned her, left her to take care of herself and the baby inside her. Feel her panic as she thinks about the minimum wage job she's had, a job she will have to quit to deliver and care for her baby. Feel her anxiety about the debts, the bus

rides she will have to take to get to the doctor's office. The
trips to social services. Feel her terror over where she will
stay, how she will provide for her child. Now, as you are
experiencing her terror, notice any thoughts that come up for
you. Just notice. Do not judge. Just allow the thoughts to
surface. Good. Good. And now take a deep breath and come
back fully into the room.

How was this exercise for you? What thoughts arose? Did you feel compassion? If you did, then what happened? Can you now be aware of your mental defenses against the feeling of compassion? Did you have religion-based judgmental thoughts? Some people doing this exercise may have thought, "Well, she should have followed the bible and this would not have happened." Others may have thought, "If she'd stayed in school she'd have a career and could take care of herself." "She made a bad choice and now she has to pay for it." Could you be aware of your own fear that you could have been the single pregnant young woman in this exercise? If you are a woman, did you ever have unprotected sex and then sit in anxiety until your next period came? If you are a man who had unprotected sex, could you be aware of your thoughts and feelings at that time about the woman you might have inseminated? What did you do when she called or texted you to tell you she was pregnant? This is not an exercise in blaming. The hope is that male or female you will come to a deeper understanding of your unconscious motivations and defenses.

Once one has broken through denial and begun the process of becoming self aware, self-awareness and Theory of Mind are sustained *because* they provide a basis for recognizing deception. One must be aware of oneself as a unique entity in order to be aware of another's mental state or deception. Recognition of deception is one difference between the progressive mind and the conservative mind. Progressives are aware of the deceptions of Monsanto regarding Round Up, they are aware of the deceptions of the Military-Industrial-

Political Complex, they are aware of Trump's deceptions, they are aware of the deceptions of the congressmen and senators who lie about the solvency of Social Security in order to destroy it.

Where deception is a viable means of gaining access to valuable resources like money, property, prestige, power or sex through competition there will be pressures to destroy Theory of Mind and destroy awareness of the self. In America deception is a viable means for gaining access to valuable resources like money and energy, and Theory of Mind and self awareness among citizens have collapsed. Even the wealthy, overcome by the Contagion do not appear to have improved Theory of Mind or self awareness. Trump appears to believe his lies.

Men are sometimes motivated by self-absorption and a desire for power and control over women to deny women pleasure in sex. Sexual selfishness driven by Contagion causes men to focus on their own satisfaction. They treat women as objects for their own gratification, not as equally deserving sexual human beings. The Contagion-infected man "gets off" then rolls over in the bed and goes to sleep leaving the woman unsatisfied. Domination and control over women reaches a high point in those African countries where men enforce clitoridectomy, cutting off the girl's clitoris when she reaches puberty. Some men are threatened by a woman's sexuality, and they try to control her sexual pleasure.

Contagion and the Poor

Contagion prevents awareness of the suffering of the poor who are denied health care, housing, access to toilets and showers, and food. The existence of poor people threatens the mortality of people who have managed to secure housing, health care, and food for themselves, whether they are middle class or wealthy. Those who have "made it" cannot allow themselves to mentally model the experience of the destitute, of those who have been so traumatized by war or life that they cannot function in consensual society. They cannot imagine having been raised in terrible family environments that produced mental illness in the children who were then unable to succeed and who may have turned to drugs or alcohol to ease the emotional pain. The Contagion affects the thought process of people who have succeeded telling them they worked hard for what they have, never recognizing the place that sheer luck played in their success. They tell themselves the poor are shiftless "takers" who don't want to work. They vote for representatives who take away food stamps, who close medical and community mental health clinics, they demand that police close homeless encampments, and they refuse to fund housing and facilities for impoverished citizens. Fear of death is activated by other people's homelessness and poverty and motivates Contagion-driven behaviors that result in denying care to people who are challenged to survive in consensual society.

Powerlessness

The Contagion damages the brain's right frontal cortex and prevents the self from recognizing its powerlessness. The fundamental powerlessness is the powerlessness over death. People feel powerless to prevent a mass murder by a psychotic person armed with an assault rifle from coming into a school and murdering their children. Powerless to stop someone else from coming in to their church, a nightclub, a mall, a movie theater, a bank and shooting random people to death. Deep in their unconscious people know that climate change is real, but they cannot allow that knowing to rise to awareness. The Contagion uses hate, denial, lying, misogyny, racism, bigotry, homophobia, and xenophobia to create an illusion of power which is a defense against the feeling of powerlessness over our mortality. People also sometimes use drugs to defend against death awareness. The first of the steps in a 12 Step recovery program states "We admitted we were powerless over drugs...." Recovery from Contagion requires the same admission of powerlessness and the same investigation of exactly how powerlessness over fear of death has manifested in our lives.

Power and the Right Wing

This section explores the issue of power and the the conservative movement known as the right wing. Individual citizens in present-day America lack power. Only the very wealthy have power. They have bought Congress and so Congress has no power. They have bought the Supreme Court so it has no autonomous power. Individual citizens have no power and that is terrifying for them, because powerlessness in daily life translates into powerlessness over death.

In order to gain access to more money, more property, and more social and political control, the powerful seek to destroy Theory of Mind in citizens so that the citizens will be less able to recognize the deceptions being foisted on them. Through buying boards of education, the powerful remove references to income disparity from textbooks, and they remove courses teaching critical thinking from high schools. They force teaching of religion in public schools. Religion facilitates the Contagion.

Religion confuses citizens by telling them everything that happens, whether good or bad for citizens, is God's plan when in fact it is the plan of the powerful that makes bad things happen. Teaching religion includes teaching creationism which separates humans from oneness with the world. When humans are separated from oneness with the physical and biological world they cannot feel the pain of the oceans when they are clotted with plastic or the pain of the atmosphere when it is poisoned with smog. When humans are separated from oneness with animals they do not feel their pain when they are shot, poisoned, or trapped. Religion allowed humans to avoid feeling the pain of the Indigenous mother or father watching as their children are murdered. Religion allows the lynch mob to avoid feeling the anguish of the African American being hung. Religion allows the NRA to avoid feeling the anguish of the mother of the murdered child shot to death by an assault rifle.

Since 2010 the conservative movement has spawned a more violent group known as the alternative right or alt-right. The alt-right embodies all of the aspects of the Contagion in a more virulent and potentially deadly form. Groups united in the alt-right include white supremacists, neo-Nazis, anti-semites and neo-fascists. Alt-right attitudes are islamophobic, homophobic, racist, misogynistic, isolationist, Christian fundamentalist, and nativist. Donald Trump's presidential campaign connected ideologically with the alt-right, and Trump brought several alt-right leaders into his cabinet after his election including Stephen Miller, Steve Bannon, and Sebastian Gorka. Trump supported the candidacies of alt-right darlings Roy Moore and Joe Arpaio. The alt-right claims to have influenced mass murderers like Adam Lanza, Dylann Roof, and James Holmes.

A goal of this section has been to expose the enormity of the Contagion's effects on society and to develop an awareness that the root cause is present in almost every segment of our culture. The alt-right is a more extreme manifestation of impulses that are ubiquitous. Most people fear death, and few people have come to terms with their mortality. Our society supports a murderous war machine that externalizes citizens' fear of death and then pays little attention to the killing it does. Some in our society who lack impulse control and emotion management skills act out violently themselves what the society as a whole acts out through agents like the military, the police, ICE, the FBI, Secret Service, CIA, and NSA. The murderous acts committed by individual citizens and by agencies are not aberrations. They express a profound unconscious schema, and they will not cease until as a society we resolve our fear of death.

Contagion and corporations and bureaucracies

Contagion prevents people from recognizing their loss of identity. Contagion promotes merging one's identity with the corporation, the bureaucracy, or the state. People who have lost their individual identity and have merged with the corporation, bureaucracy, or state are unaware of the shift in identity. Loss of individual identity serves the interests of the larger unit, and those entities work to supplant the identity of the self with the identity of the corporation. In America this programmed transformation reached a zenith in the indoctrination programs at IBM. An IBM songbook contained words and music for "Ever Onward: The IBM Rally Song." The song celebrated Thomas Watson, IBM's founder, and as employees sang it, their identities merged with the corporate identity. "Thomas Watson is our inspiration/Head and soul of our splendid I.B.M./We are pledged to him in every nation,/Our President and most beloved man."

When individuals assume the identity of the corporate entity, they lose the human values of compassion, care, loving kindness, and empathy, qualities of the individual, qualities that the corporation does not possess. The corporation, bureaucracy, and state have succeeded precisely because they have eliminated the prosocial qualities of the individual. When citizens infected with the Contagion work for Northup-Gruman, General Dynamics, Raytheon or any of the corporations of Big Defense, they lose the Theory of Mind that would otherwise allow them to feel anguish at the murder of children by the bombs and planes and rockets they manufacture. The same holds true for employees of the Defense Department, whether the U.S. Defense Department or the similar agencies in other countries. A primary objective of indoctrination into the military in "boot camp" is to eradicate the individual's identity so he or she will feel no compassion or remorse about killing other human beings.

When humans identify with the corporate or bureaucratic entity they lose the ability to feel compassion for the atmosphere contaminated by the coal companies just as they lose the capacity to feel compassion for the humans poisoned by the contaminated air. They lose the ability to feel empathy for the land polluted by fracking companies pumping waste into the water wells. The also lose the ability to feel terror that they are contaminating their own drinking water. People act as if they believe that corporations will survive and so will individuals because they are merged with the corporation. When the people are poisoned or die of starvation or thirst and there are no more people to work in the government or to fight for the military or to work for the corporation, there will be no more corporations or bureaucracies.

We human beings have been conditioned by aeons of evolution to participate in tribal societies where the central nexus is the family, then the clan, then the tribe. Survival depended upon membership in a relatively small unit of people. The Contagion has damaged the awareness that we no longer live in a tribal society. However, people's unconscious schemas motivate them to experience the same emotional attachment to the corporation, bureaucracy, or state that they once had for the family, clan, and tribe. People are used to identifying with tribes like the borough or city they grew up in, the school or college they attended, their fraternity or sorority, their unions or professional associations. Ford, IBM, General Motors, Exxon, Bank of America are not tribes but the loss of Theory of Mind caused by the Contagion makes it difficult to recognize that fact. People used to relate to the major political parties as if they were tribes, but they are not. The major political parties are bureaucracies without compassion or empathy, manifesting the properties of corporations, serving the interests of the donors, not the needs of ordinary citizens. Like a bureaucracy, the Republicans and Corporate Democrats deceive the citizens and spread lies about their opponents to secure the donations and votes they need to maintain power. In the contemporary era Republicans have promised tax relief for the middle class, but their tax bill gave trillions to the donor class and almost nothing to the middle class. Yet citizens hold on to affiliation with Republicans as if their lives depended on it. Certainly their identities do depend upon affiliation with the Republicans, and such is the power of unconscious schemas, especially the death anxiety.

Chapter Four

Hospice for Our Home World

We've developed a model for the Contagion that has infected our species, and we've traced the causes of the Contagion to our very human death anxiety, our terror at the prospect of our own mortality. In this chapter we'll amass the evidence that the world is indeed dying. The chapter will argue that the window of opportunity to reverse the course of our world's death has closed. The message of this chapter derives from studies of patients who are in fact dying. When a patient has contracted cancer or some other fatal disease and he or she is approaching death, the attending physician will assemble the family and inform them of the patient's condition and the likelihood that further medical interventions might prolong life or not. The patient may have already signed an Advance Directive expressing his or her wishes about how they want to go about dying. The family will be informed that the death process has begun, and the dying patient will be moved from hospital to hospice where medical efforts shift from curing the patient to helping the patient transition out of life. The patient and family will be helped by hospice staff to say good-bye to each other.

As you will read, the doctors—in this case the climatologists and geologists and environmentalists and zoologists and psychologists—who have cared for the world are giving their assessment of the world's disease's progression and the chances that any intervention could reverse its course. The doctors have come to the patient's beside with somber faces. Their assessment is that the Contagion—what Levy calls by its Indigenous People's name, *Wetiko*—has spread too widely through all the systems of the world. As you read this chapter, allow yourself to think about what would be required, the massive changes that would have to happen if we were to try and reverse the course of the Contagion that is eating at our world. Allow yourself to be aware of the behaviors that would have to change and change collectively, not just in yourself or your family, but in every human being in order to reverse the course the world's death process. Consider the disordered emotions, thoughts, and attitudes that would have to change if we humans were to mount a truly heroic intervention. The patient dying of cancer and his family always hope that a revolutionary cure will suddenly appear to reverse the course of the cancer.

What revolutionary cure would that be that could stop the Contagion? It's like asking a morbidly obese person if he or she is capable of sticking to a regimen of diet and exercise and psychotherapy necessary to lose 300 pounds. As you read ask yourself if you think humanity could muster the will to rein in the corporations, to stop the military, to change the consumerist identity of civilization, to break the addiction to mindless expansion, to reduce the carbon footprint of every single one of us, to break through the greed and self-absorption and begin to work together toward a common goal, the goal of survival for our world and ourselves. Ask if you think we can as a species forgo reproducing, to discuss overpopulation, and to voluntarily accept strict limits on family size, for example, two children only. Do you think we can do this?

Powerful forces are arrayed that oppose any changes as radical as those necessary. It appears the Contagion has infected humanity too deeply, and there is no will to effect the changes necessary to stop the death process. The existence of corporations and bureaucracies depends upon their not changing their behaviors, and so they are not going to take the actions necessary to save our world. Big Energy corporations continue to act as if climate change is a hoax perpetrated by socialists or liberals. They deny the science that proves climate change is happening, and they have bought the elections of Congresspeople and a President who adhere to that view. While the environment deteriorates by daily measures, Big Energy and a complicit Congress continue to frack and to expand drilling into pristine wilderness. Besides the corporations and bureaucracies, humans' thinking, feeling, and acting are too badly compromised by the Contagion. Their denial is too great.

The billions of people who live on Earth have no other means to continue living besides continuing with the very behaviors that have produced climate change and the degradation of the environment and the deaths of the plants and animals who live here. People who have jobs have no way of supporting themselves or their families except by continuing to work at their jobs, and so they will persist in commuting by car, driving to the store, driving or flying on vacations all the while feeling a deep dread that they cannot allow to rise to awareness and all the while hoping that someone will come up with a plan.

No one is going to come up with a plan. One hundred ninety-five nations tried to come up with a plan, the Paris Agreement, that went into effect November 4, 2016, whose goal was to combat climate change by pursuing efforts to keep the level of global warming below 1.5 degrees Celsius. The main action step proposed by the treaty was for signatories to report their levels of emissions. The treaty recognized the critical importance of keeping warming below 2 degrees Celsius. Donald Trump was inaugurated President of the United States on January 20, 2017 and he lost little time promoting his America First program. On June 1, 2017, he announced that the United States would cease participation in the treaty. He said that the treaty "undermines the U.S. economy" and that it "puts the U.S. at a permanent disadvantage." Trump asserted the Contagion as a reason to withdraw from a treaty that was taking a small step toward curing the Contagion. It was as if a cancer patient, on a day when he was feeling better, suddenly pulled the IV needles from his arm and announced he was stopping the treatment because it was weakening him. Nations of the world reacted with dismay, however they continued to trade with the United

States and to collaborate on military and economic projects around the world. The Republican party celebrated the President's bold action. Corporations and bureaucracies rewarded themselves for having avoided a plan that might have affected their bottom line. Contagion-driven increases in all the metrics of global warming continued unabated. CO_2 levels, methane levels, melting polar ice, frequency and intensity of hurricanes, heat waves, and global temperature increases are symptoms of the problem. The cause of these symptoms is the Contagion and the ways of living that the Contagion has caused, and no one is addressing the profound restructuring of how humans live their lives that will be necessary if the world is to survive.

The argument of this chapter is that the necessary change will not happen. The world is doomed. The world is headed for the hospice. At some point in a cancer-infected individual human's life, the physician takes the family aside and explains that the disease has metastasized too widely in the body. There will be no cure. The family must prepare for the patient's death, and the patient must go to hospice to make the transition as painless as possible. Physicians usually prescribe opiates for patients undergoing the death process to ease their physical and emotional pain. America in 2017 is in the midst of an epidemic of opiate abuse, and we can ask ourselves to what degree are people self-medicating the anguish they feel as the death of our world begins to rise to awareness.

The following paragraphs will summarize the doctors' observations of the world upon which they are making their assessment that the patient is moribund and will not survive. The goal is the same as the doctors' goal at the cancer patient's bedside: to convince the family that further heroic efforts will be useless. These will be difficult paragraphs to read, just as it is difficult for family members to hear that their beloved relative is dying. In the following chapter we will speak about what we can do for ourselves and our world as we enter hospice.

An excellent and comprehensive review, "Uninhabitable Earth," summarizes the effects of climate change. I have used this review to organize the material in this chapter. The review appeared in New York Magazine on July 10, 2017 and was written by David Wallace-Wells.

Melting permafrost and release of methane

The Intergovernmental Panel on Climate Change, IPCC, has reported that daily maximum and minimum temperatures in the Arctic are 2 degrees Celsius above average. Temperatures in the region are currently higher than they were at 44,000 to 120,000 years ago. The rise in temperature is caused by increased anthropogenic greenhouse gases and deposition of soot on the ice. As the ice melts, the albedo (reflectivity) of the poles decreases exposing dark land and dark water below. The dark land and dark water absorb more heat, which accelerates the warming process. As the Arctic warms the permafrost melts. Permafrost is soil including rocks and water that has remained below the freezing point of water, 0 degrees Celsius. The layer of permafrost varies in thickness and may have been permanently frozen for eons. When permafrost warms above freezing, carbon compounds that have been held in the frozen soil for tens of thousands of years are released. The carbon compounds were dissolved in the soil and consist of CO_2 and low molecular weight organic acids that are quickly converted to CO_2 and escape into the atmosphere. In addition to CO_2, methane, CH_4, is also trapped in the permafrost and released when it melts. Geochemists estimate that 1.8 trillion tons of carbon in the form of carbon dioxide and methane are trapped in the permafrost.

The atmosphere holds 0.7 trillion (0.7×10^{12}) tons of carbon in the form of carbon dioxide right now which makes up 0.041% by volume of the atmosphere. The amount of carbon trapped in the permafrost in the form of carbon dioxide and methane is about twice the amount of carbon already present in the atmosphere. The amount of carbon in the atmosphere right now is causing the temperature of the planet to increase. Climatologists have agreed on a baseline for the Earth's temperature as the average from 1951 to 1980. During that period the Earth's average surface temperature was 14 degrees Celsius. The Earth's current surface temperature is 14.9 degrees Celsius which means we've warmed up 0.9 degrees, All of the effects cataloged in this chapter are the result of that 0.9 degree Celsius increase. Climatologists advise that we avoid reaching a 1.5 degrees Celsius increase, the goal of the Paris Accords. The Paris Accords hoped to hold the temperature increase to no more that 4 degrees Celsius, but the consensus now is that the temperature will increase 4 degrees Celsius by 2200 and could be as high as 8 degrees Celsius. In a rare display of candor, the Trump administration recently estimated the rise will be 7 degrees Celsius.

As the planet warms and the permafrost and the Arctic ice melt, more and more methane is released into the atmosphere. Methane is held in an icy slush on the floor of the ocean where the water pressure keeps in from bubbling up to the surface. As the ice melts the pressure decreases and the gas rises up. Methane is a more potent green house gas than carbon dioxide. On the timescale of a century, methane is 34 times more potent than carbon dioxide. On the time scale of 20 years it is 86 times more potent. Scientists estimate that the progression towards at least a 4 degree Celsius temperature rise is irreversible because of the carbon already in the atmosphere which is being added at the rate of 40 billion tons per year. The cancer that is killing our world has metastasized to the point where there is really nothing we can do to stop it. We have entered the Sixth Mass Extinction. Like four of the previous five mass extinctions, this one is caused my warming and release of methane. Only one of the previous five was caused by an asteroid, the *Chicxulub* asteroid that hit the Earth 65 million years ago and resulted in the death of the dinosaurs and much other life on the planet. Human beings and our activities are the cause of the Sixth Extinction.

As the world warms

We humans and our mammalian relatives evolved in a planet with a narrow range of temperatures. The ideal outside temperature for human beings is a small range from 15 degrees Celsius to 25 degrees Celsius. Humans must maintain a body temperature of 37 degrees Celsius, and they will die if the body temperature decreases to 21 degrees Celsius. Humans cannot survive in an outside temperature of 43.3 degrees Celsius. The lethality of the temperature depends upon the humidity. A combination of heat and humidity determines survivability. When the ambient air is dry, humans can survive higher temperatures, but when the water content of the air (the humidity) increases the body cannot cool itself because it cannot dispose of excess heat through evaporation. Wet bulb temperature measurements determine the interaction of humidity and temperature. Wet-bulb temperature is measured with a thermometer whose bulb is wrapped in water-soaked cloth. Humans can survive for only a few hours if the air temperature is 35 degrees Celsius and the humidity is above 90 percent. The higher the temperature the lower the humidity that humans can tolerate. At an air temperature of 40.6 degrees Celsius and a humidity of 90 percent, humans will cook from both inside and outside and will die within a few hours. As the world warms, more and more moisture evaporates from the ocean, and both the temperature *and* humidity increase in cities on the coasts.

The image of the frog in a pot of water slowly being brought to a boil is apt for human beings' situation on Earth right now. The warming of the planet is insidiously slow enough that most people are not alarmed by it. That is unless one lives in India or southern Pakistan. In May, 2015 a heat wave killed 2,500 people in India and a month later in June the heat killed 2,000 in Pakistan. Temperatures rose as high as 48.9 degrees Celsius. Zoo animals and livestock were also killed by the heat. In 2003 70,000 Europeans died during a heat wave that lasted two months. Denial of our own mortality tells us these extreme heat events are isolated instances happening far away. Our fear of our own death prevents us from awareness that extreme heat events are becoming steadily more frequent. In fact, the frequency of extreme heat events has increased 50-fold since 1980. Examining records dating back to 1500 the five hottest summers have taken place since 2002. As the planet warms, the extreme heat events persist for longer and longer periods. The 2003 European heat wave lasted over two months that summer and resulted in sustained temperatures of 40 degrees Celsius. The elderly living alone were most susceptible. When the global temperature rises to four degrees Celsius above the historical average, events like the European 2003 heat wave will become the norm. At six degrees above the historical average, New York will be hotter than the nation of Bahrain at the edge of the Sahara is today. In the Mississippi delta, people would be unable to work, and humans would die in their sleep from hyperthermia. Attempts to lower the heat with air conditioning will worsen the situation as more and more carbon is added to the atmosphere from coal-fired electrical plants.

A study released at the end of October, 2018 disclosed that oceanographers using improved methods for measuring the amount of heat the oceans have absorbed found that the oceans world-wide have absorbed 60 percent more heat than previously believed. The oceans are an enormous heat sink. Heat absorbed by the ocean's waters is transferred to the atmosphere above the ocean, contributing to increasing the destructive power of hurricanes. Warmer oceans also melt Arctic shelf ice from below, thereby increasing the rise in the height of the oceans. As science improves the accuracy of its measuring techniques we are learning that the effects of climate change are much more dire than previously believed. Contagion prevents transfer of this knowledge from the scientific journals into the American consciousness. Contagion prevents citizens from awareness of the knowledge because it triggers mortality anxiety.

At a time in human history when our species most needs informed leadership, denial of awareness of our impending doom has reached ridiculous levels. Contagion thinking affects all classes of society. Oklahoma Senator and notorious climate change denier James Inhofe brought a snowball to the U.S. Senate in February 2015 to "prove" that global warming was a hoax. The year 2014 had already been declared the warmest year on record. Inhofe could have used the snowball to demonstrate that temperatures fluctuate wildly as they trend ever higher. So-called leaders are often the most infected with the Contagion. Money and greed contribute to the Contagion, and Senator Inhofe has powerful monetary incentives to deny the existence of climate change. In the period from 1989 to 2018 Inhofe received $1,883,610 in donations from oil and gas industries most notably from the Big Energy corporations, Koch Industries and Murray Energy.

Global warming and Global climate change are two of several measures by which the climatologists are concluding that our world is dying and needs to go to hospice.

Food scarcer in a warmer world

Cereal crops like wheat, rice, oats, barley, and corn are staples of human diets world-wide. As the planet warms, crop yields decrease about 10 percent for every degree rise in temperature. The decrease could be as large as 17 percent. While climate change decreases food supplies, the demand for food increases because the population continues to rise world wide. As of October, 2018 there are 7.7 billion human beings demanding to be fed on planet Earth. Even without climate change the planet is not capable of feeding all of these people. At current consumption, Earth could support 2 billion human beings. If the people living today consumed resources the way the United States does, 4.1 Earths would be required to sustain the lives of the 7.7 billion. We do not have 4.1 Earths. We only have the one.

Food depends on water, and there is less and less of it available. California is a good example. A great aquifer lies beneath California's central valley and industrial farmers and cities have been pumping the aquifer so fast that the land above it is sinking. The Ogallala Aquifer located beneath the central United States, like the central valley aquifer in California, is being pumped dry, and scientists estimate it will go dry in 50 years. Scarcity of water is leading to conflicts between industrial farmers and local resident farmers. Because the Contagion has affected people who own and work for industrial farming corporations, greed has overwhelmed any compassion they might have for people whose wells have run dry because the industrial farms have sunk deeper wells and are emptying the aquifer. Water departments and politicians have been co-opted by the corporations, the best lawyers money can buy have rewritten the laws regulating water use, and the corporations are winning the war for water against ordinary citizens.

As climate change alters weather patterns, rain fall decreases, water becomes scarcer, and drought becomes more prevalent. Food crops cannot grow in drought conditions. The U.S. Southwest was already in an extended drought in the spring of 2018. The areas of drought spread into Oklahoma, Texas, and Kansas. Pockets of drought appeared in Arizona and New Mexico. Exceptional drought dried up the Oklahoma panhandle and pushed out into Utah and Kansas. Parts of seven states in the U.S. Southwest were parched by exceptional drought conditions. In spite of one wet winter in the 2016-2017 rainfall year, California has experienced several years of drought conditions. While the entire U.S. appears to be drying out, the federal government and politicians have not mentioned drought, dwindling water supplies, and changes in rainfall. The Contagion prevents politicians from speaking out about drought, apparently out of fear of upsetting citizens and wealthy donors. The potential for catastrophe grows as these effects of climate change develop outside the public's awareness.

A NASA study predicted that the western half of the U.S. including central and southwest U.S. will experience a drought many times worse than the dustbowl of the 1920s if greenhouse gases continue to accumulate. This drought will last 20 to 40 years and will set in motion another great migration just as the dustbowl did. In spite of this clear and present danger, the government continues to focus on the frenzied activities of the current president who distracts citizens by focusing on Hillary Clinton's emails. The Contagion is preventing the government and citizens from dealing with an imminent threat to human survival. This is predictable behavior, because knowing that death is coming, and not just individual death but the death of the world, people are faced with their mortality, but because they have never resolved the issues of death in the past, they are ill-prepared to start now. It's as if the family of the dying person is gathered in the hospital room but instead of talking to each other or to the soon-to-be deceased, they are watching TV.

Drought is a world wide phenomenon. Drought has been ongoing in southern Europe as rainfalls have dropped by 50 to 75% below averages. In November, 2017 devastating drought set in and rivers ran dry, crops failed, and wildfires broke out in the aftermath of the drought. NASA photographs from space show vegetation has thinned. Italian reservoirs are at 50% of capacity. The Middle East nations around the eastern end of the Mediterranean have suffered as well. Drought began in the Middle East in 1998, and scientists studying tree rings conclude the current drought is the worst in 900 years. From Greece drought spread through Bulgaria, Romania, east into Ukraine, then down into Syria, Lebanon, and into Egypt. Africa is not immune. The United Nations estimates that 20 million could die from starvation caused by drought in Somalia, South Sudan, Nigeria, and Yemen.

Drought is having a powerful effect on Australia. In a country that has historically experienced drought once every 18 years, the so-called "millennium drought" parched the continent between 1996 and 2010. Patchy rain in 2003 and 2004 failed to break the drought, and when rain did come in 2011 if often brought flooding although it did end the drought for a few years. By 2013 to 2015 drought again began to develop, and as a result of dry conditions a "gigafire" broke out that consumed over 2 million acres of brush. There was scant water available to stop the conflagration. The Contagion is a powerful force in Australia as elsewhere, and in 2018 the government of Scott Morrison is spending its energy on fracking and building pipelines which assuage the corporate forces even as it furthers destruction of the environment.

China has experienced a drought that threatens the food supply of its 1.4 billion people. The drought began in the fall of 2010 when rain and snow failed to fall on the southeast provinces. Major lakes dried up. The wheat harvest failed leading to a world wide shortage of the grain staple. Ultimately 35 million Chinese people lacked adequate drinking water and in some locations people were forced to relocate.

Americans think that these countries are thousands of miles away and the events there cannot happen here because they have not happened in the life spans of most people alive today. Such is the power of the Contagion to disrupt thinking and feeling. Such is the power of mortality anxiety.

Fire

As the planet warms and drought spreads, the world's forests and grasslands become more and more susceptible to catastrophic fire events. Portugal, Spain, and Greece illustrate how fire follows drought. We've already recounted how devastating drought parched southern Europe reaching a peak in 2017. After drenching rains that produced flooding, the drought seemed to have abated, but it returned with a vengeance to desiccate Europe from Sweden to Greece in 2018. Fires returned to Portugal, Spain, and Greece, and scores of people were killed.

Fires have intensified in number and extent in the United States. During the 10 year period ending in 2005, Texas experienced 160,000 fires that burned 9,489,000 acres. In 2011 alone 1,969 fires ravaged Arizona and burned over a million acres of wild land. The fires in the American southwest continue. As of May, 2018 824 Arizona fires had incinerated 75,000 acres. In California climate-change caused drought has worsened the number of fires and the destruction caused by fire. During just the first 9 months of 2018, California was hit by 6,124 fires that burned an area of 1,470,483 acres. For comparison, in all of 1980 when climate change was beginning to worsen, California reported 69 fires that burned 101,803 acres.

Fire worsened by drought no longer respects city limits. In October, 2017 a fire began east of the city of Santa Rosa. It spread quickly in brush and trees parched to tinder dryness and driven by winds that increased to 40 mph. By the time it had had its way, the so-called Tubbs Fire had charred parts of three Northern California counties: Napa, Sonoma, and Lake. The Tubbs fire laid waste to 37,000 acres, destroyed 2,800 homes, motels, and other structures worth $1.2 billion, and immolated 22 people. The fire occurred in an area called Fountaingrove that had experienced devastating fires three times previously in its history. The earliest fire to consume the Fountaingrove area occurred in 1870 before conflagrations were named. It seared the hills and canyons running east and west from Calistoga where it began, incinerating everything in its path until it reached Santa Rosa. The Commandery fire of 1908 blackened the same hillsides and destroyed the homes that had been built there after 1870. The Coffey Park fire of 1939 destroyed the same area that was consumed by flames 78 years later in 2017.

In a stunning testament to the hold the Contagion and the denial of death have on people, after each deadly fire residents rebuilt in the very path the previous fires had taken. Even today in 2018, luxury homes are being built on the ashes of the residences destroyed just a year previously. Within two months of the end of the Tubbs fire, some of the 1,500 burned out properties began to appear on realtors' web pages. Showing the grip that the money, greed, and property components of the Contagion have on people, realtors were asking from $140,000 to $800,000 for scorched barren land that had burned four times in the previous 147 years.

To more effectively fight these kinds of fires, the California Department of Fire and Forestry and Protection has pleaded for aerial supertankers built on the Boeing 747 frame. An aerial fire engine, it can carry 19,600 gallons of fire retardant. The planes cost $120,000 per day to operate. The federal government is limiting the role played by aerial supertankers. In an example of the Contagion in action, a private company received the contract to send the much smaller DC-10 to fight forest fires. The needs of people and forests were ignored in order to pay money and satisfy the greed of the apparently politically connected DC-10 provider. Meanwhile, the Contagion-driven Congress and Executive have secured a military budget of $770 billion to fight land wars in the Middle East and Afghanistan while America is consumed by flames. Tax dollars that could have been spent on fire-fighting planes to protect the homeland are being invested in planes to drop bombs on foreign countries.

Climate change has brought wildfires inside the Arctic circle. Russia, Finland, Norway, Sweden, Canada, and Alaska are all experiencing increased numbers of wildfires in 2018 that are incinerating larger and larger amounts of forest when compared with previous years. Devastation from wildfires has been exacerbated by the climate change related spread of bark beetles. Before global warming accelerated in the 1980's the winter freeze in the Arctic circle kept the beetles in check.

Now as the climate warms, bark beetles are able to cram two reproductive cycles into the late spring and summer months enabling a population explosion that is devastating the pine and spruce forests from the Arctic down into Washington, Oregon, California, and Colorado as well as into Sweden where thousands of acres of trees have been killed. Trees killed by bark beetles are kindling for forest fires.

The situation in the Matto Grosso, the Amazon Rainforest, dramatically demonstrates the interaction between Contagion and fire. The Amazon rainforest is being logged and burned to ranch cattle for export to serve the American and other world markets for beef for burgers and soybeans for export. Large industrial farms growing soybeans have replaced the forest that had stood for immeasurable centuries supplying about 20 percent of the oxygen we breathe. Areas of forest have been cleared for dams, to build settlements for poor farmers who work the fields, for roads, and for timber. The soil of the Amazon is notoriously poor farm land because the soil is thin and nutrients are stored in vegetation. Burning does not return nutrients to the soil. In addition to industrial farming, slash and burn agriculture has also been practiced in the Amazon, meaning that an area is cleared of trees, which are burned, a crop is planted for a couple of years, and then the farmer moves on to clear and plant another area of the forest. In Brazil, Colombia, Bolivia, and Venezuela, the principal nations claiming the Amazon basin, 290,000 square miles of forest have been cleared. Studies have demonstrated that 85 years are required for the forest above ground to recover after it has been ravaged for agriculture. The soil itself does not recover nearly as quickly, and species of animals that were displaced by the slash and burn had not recovered over the duration of the study. Contagion-driven greed and Contagion-driven pursuit of money drive the deforestation of the Matto Grosso. Land speculators invade the Amazon and drive up the price of land hoping to "make a killing." Contagion prevents humans from awareness of the damage they are doing. When indigenous people complain their leaders are often murdered. The rainforest is a living being, much greater than the individual trees and other plants and animals comprising it. When speculators speak of "making a killing" they are speaking truth for they are participating in murdering a living being, and furthermore they are disrupting a reciprocal relationship among all animals, including humans, and the plants that sustain life.

Species extinction

Insatiable greed and unbounded self-absorption drive the expansion of the human footprint on the land, and as human populations expand, the populations of other animals, vertebrates, mollusks, and arthropods, decrease. The World Wildlife Foundation released a report in late October 2018 stating that overall human activities have caused a 60 percent decline in vertebrate species planet-wide in the 44 years from 1970 to 2014. Comparing current species loss to the archaeological record shows that only during the past five mass extinctions has this much biodiversity been lost. Just one-quarter of the Earth's surface is currently free from the impact of human activity, and by 2050 only 10 percent will be free. The United States and Canada have the largest human consumption footprint. South and Central America have eliminated 89 percent of their vertebrate populations. The vertebrate species include fish, large and small mammals, reptiles, frogs, and birds. Frogs, toads, and salamanders are disappearing from their habitats and could be gone completely in six to 20 years. Invertebrate species are also in decline due to habitat destruction and agriculture. A western European study at over 100 nature reserves reported an 80 percent decline in flying insects. Honey bees, monarch butterflies, lightning bugs, moths, hover flies, dragon flies, and beetles all showed massive losses. The authors note that flying insects form the basis of the food chain for birds, so when the insects disappear, the birds begin to starve. Another class of invertebrates, the mollusks like snails, clams, and mussels are dying off. In one of the rare studies, up to 50 percent of freshwater mussels died of over two years. The causes are habitat degradation, disease, and pesticide and herbicide spraying as well as chemical spills. Planetary health depends on there being a large number of diverse species in the biosphere. Monoculture is ruinous. Earth is moving in the direction of a monoculture of human beings as we eliminate other species.

Choking on the air we breathe

Pollution of the atmosphere has spread in lockstep with the spread of industrialization. By the 1800s, the usual fogs of London had become deadly due to addition of coal smoke to the mix that was named "smog" by Dr. H.A. des Voeux in 1905. The situation became catastrophic in 1952 when thick, yellow, acidic smog blanketed London for five days and choked 4,000 to death immediately, made 100,000 sick, and eventually killed an additional 6,000 in the months following the event. No long term studies have measured the changes in life expectancy due to Chronic Pulmonary Obstructive Disease and lung cancer that would be expected outcomes from the 1952 London smog event. In the 66 years since the "Great Smog," scientists have amassed a formidable amount of data proving the enormous danger to human health caused by air pollution.

Air pollution kills 4.6 million humans per year. The World Health Organization states that 2 billion children live in atmospheric pollution that exceeds international limits, and that 300 million are trying to breathe in air that is six times the international limits. The vulnerable populations include the elderly, people with heart or lung disease, and people with diabetes. Air pollution increases incidences of asthma, cardiovascular disease, Chronic Obstructive Pulmonary Disease, diabetes, and cancer. Exposure to polluted air lowers IQ and increases risk of autism. A definitive study that appeared in December 2019 demonstrated that breathing polluted air shortens one's lifespan by 2.6 years.

Multiple sources contribute to atmospheric pollution including industrial emissions, automobile exhaust, and wildfires. The forest fires of 2017 and 2018 gave Californians a lesson in atmospheric pollution and how widely particulates from fires can circulate. San Francisco Bay Area residents watched the dirty clouds roil down from fires in the north and spread out over the central valley. The smoke plumes from the Mendocino Complex and the Carr fires could be seen from the space station orbiting the Earth, and the particulates traveled as far as the east coast. Smoke from the Tubbs fire in 2017 was so dense that people wore masks. Particulates in the size range of 2.5 micrometers—smaller than pollen—can penetrate deep into the lungs, and scientific studies show that these particles can cause lung and heart disease. At one time in America, smog from burning coal and from car and truck exhaust was severe, and Los Angeles and New York looked then much like pictures seen today from Peking and Shanghai in China. The Clean Air Act of 1970 dramatically reduced the levels of air pollution, and children have been safe to play outdoors during school recess once again. Infected by the Contagion, President Trump and his former E.P.A. chief Scott Pruitt promoted an activist agenda that included loosening emission levels and fuel economy standards. In a Contagion-driven pattern, Pruitt had received contributions from Big Energy. He asserted that E.P.A. standards were motivated by an anti-business agenda that was too restrictive. The Contagion placed the demands of business above the needs of humans to breathe clean air. Americans acquiesced without complaint because their ability to recognize deception had been impaired by the Contagion. Americans also felt powerless to do anything about the foul air they were being forced to breathe. Pruitt eventually was forced to resign over ethical violations, not because of his egregious policies.

Combustion of coal, natural gas, and gasoline produces carbon dioxide, CO_2. Most organisms that break down carbon compounds for energy produce the gas. We humans and other animals breathe in oxygen which is consumed in extracting energy from sugars and other foods, and the oxygen is combined with carbon to make the CO_2 we exhale. Burning coal, natural gas, and gasoline produces far more CO_2 than does human and animal respiration. Carbon dioxide is a greenhouse gas, meaning it traps infrared light. Sunlight can penetrate an atmosphere containing CO_2, but radiant energy—infrared light—from earth warmed by sunlight is less able to escape. This is the cause of global warming.

For at least 400,000 years, the CO_2 in the atmosphere fluctuated around 280 parts per million. Beginning with the start of the Industrial Revolution, atmospheric CO_2 began its inexorable rise, and the temperature of the planet began to rise with it. Currently the CO_2 level stands at 410 parts per million (ppm), and in worst case scenarios it will reach 1,000 ppm. In addition to contributing to global warming, this high a concentration of CO_2 reduces brain functioning by about 20 percent. Therefore, at a time in our history when humans need to think more clearly than ever before in order to find solutions to the problems we face, our brain function is impaired by the very gas that is causing the planet to warm.

Carbon dioxide is not the only gas that endangers survival of life on Earth. Ozone is formed in the upper atmosphere when ultraviolet light strikes an oxygen molecule, O_2, causing release of an energetic atom of oxygen. That highly charged atom quickly reacts with another oxygen molecule to form ozone, O_3. High in the upper reaches of the atmosphere, ozone is beneficial because it absorbs ultraviolet light that would harm humans, other animals, and plants on the surface. Ozone is also formed by combustion in car and truck engines. This is the harmful ozone. It shortens life spans and increases a child's chance of acquiring autism. Globally, by the end of the twenty-first century, about 30 percent of humans will be breathing ozone-polluted air.

What lies beneath the ice?

Biologists are concerned that pathogens have lain dormant beneath the ice, preserved in the permafrost. Out of circulation, these bacteria and viruses have been cryopreserved. Some of them may have been in deep freeze since before human beings appeared on the scene meaning we would have no immunity to them. Strains of the pandemic flu that killed 100 million humans in 1918, strains of bubonic plague and small pox, organisms not seen on the surface in decades and in some cases centuries may well be preserved in the permafrost just waiting to be released. For example, anthrax from a reindeer carcass killed 75 years earlier and held in the permafrost deep freeze killed a boy near the Siberian city of Salekhard when climate change warming melted the permafrost and the deadly spores were released. Release of permafrost-preserved pathogens comprises another potential threat from climate change.

Diseases of concern

In the Contagion-driven quest for money, property, and power, people are impelled to travel widely and diseases travel with them. In 2014, a man who had visited relatives in Africa's Liberia returned to his home in Texas carrying the deadly hemorrhagic fever known as Ebola. Ebola is highly contagious. The man died. A nurse who had treated the man fell ill with the disease. At the present moment in late 2018 an outbreak of Ebola in the Democratic Republic of Congo has killed 200, and due to economically driven warfare, physicians and health care workers are unable to reach the isolated villages to try and stop the spread of the disease.

Zika virus, which is carried by mosquitoes, suddenly appeared in the United States in 2015 apparently in a mutated form because it now causes the birth defect of microencephaly in babies infected in utero. The disease causes a moderate flu in adults. From 2015 to the present, the U.S. has experienced 5,734 cases of Zika virus infection, most of them occurring in people returning from affected areas. Brazil is the country most affected by Zika with 71,000 cases identified in Rio de Janeiro province in 2016. As with Ebola, air travel has facilitated the dispersal of Zika. Dengue and malaria, both spread by mosquitoes, have been tropical diseases that were rarely found in temperate zones. As the planet warms dengue and malaria will expand their footprint, and the World Bank expects these diseases to affect 70 percent of the world's population.

The use of quarantine to restrict the spread of plague originated in Croatia in 1377. Plague was spread by sailors, rats, and cargo arriving in Europe from the Middle East. Officials began to track the itineraries of travelers and quarantined people who'd been in areas where plague had broken out. Public health officials used quarantine most recently in 2003 to effectively stop the spread of SARS, Severe Acute Respiratory Syndrome. Today officials monitor travelers' movements and can quarantine them for monitoring if they're entering the U.S. from nations where deadly infections are active.

Melting glaciers and polar ice caps

Driven by the Contagion executives at shipping companies and Big Energy corporations celebrated this summer when the Arctic ice melted that had made ocean travel impossible from Europe to North America through the long-sought northern passage. A Russian tanker for once unaccompanied by an ice breaker immediately set sail through the waterway. Countries of the world are competing to claim the now-open sea for oil drilling. Greed for money and profit has clouded the awareness of these executives. The Contagion prevents them from recognizing or caring that profit results solely from destruction of the Arctic environment.

As the Arctic ice has thinned because of warm air from above and melted due to upwelling of warm water from below, the complex interaction of the polar ice caps and Earth's global weather patterns are exposed. The white Arctic ice reflected sunlight, an effect called "albedo." Without the reflecting ice, sunlight is absorbed and further heats the ocean water. Without the ice to cool the air above it, global air currents have been disrupted, and as a result storms are more severe with more intense winds and lasting longer. Hurricanes are more frequent and more deadly. A government report states "There has been a substantial increase in most measures of Atlantic hurricane activity since the early 1980s, the period during which high quality satellite data are available. These include measures of intensity, frequency, and duration as well as the number of strongest (Category 4 and 5) storms."

Arctic sea ice is retracting as well as thinning. In the 1980s the mean extent of Arctic sea ice was 7.2 million square kilometers as calculated from satellite measurements. By 2012 the sea ice had withered to 3.3 million square kilometers, meaning the Arctic had lost more than half of its sea ice cover. During this period the Arctic sea ice lost much of its volume as well as its extent. For eons the ice thickness had been stable at 16 feet. Now it has thinned to from zero thickness to a half a meter thickness (about 20 inches). The thinner sea ice allowed a Russian tanker to sail across the top of Russia and into the Bering Sea without the aid of an icebreaker. The Danish container shipping company Maersk sent a ship loaded with 3,600 shipping containers from Vladivostok through the Bering Sea to St. Petersburg. Contagion-infected shipping executives are giddy with joy to at last be realizing the dream of a northwest passage. Unfortunately the planet had to warm by almost 2 degrees Celsius in order for them to be able to plow their ships around the top of the world, droughts and hurricanes worsened, wild fires ravaged wood- and grasslands, and hundreds of thousands of human beings died. Such is the power of the Contagion to cloud men's minds.

As well as the Arctic ice cap, the gigantic sheets of ice covering Greenland and Antarctica are melting. Geoscientists call Earth's frozen masses of water the cryosphere, and 8 percent of the mass of ice in the cryosphere is located in Greenland. Greenland's ice mass is a mile thick in some places, and it is melting and has been since 1998. It's melting faster than it has in 400 years. The world's oceans would rise by 20 feet if all the ice in Greenland were to melt, inundating lower Manhattan and flooding the National Mall in Washington, DC. Using cores of ice pulled out of the Western Ice Sheet the scientists discovered that 270 billion tons of ice are melting off the Greenland ice sheet every year. Greenland's ice sheet is melting due to warm air stalling over Greenland as warm water circulates in the ocean and the climate warms. Greenland has warmed by 1.2 degrees Celsius due to human activity releasing greenhouse gases. The ice sheets in Greenland and Antarctica rest on solid ground, and they are held in place by glaciers at the margins of the ice sheets. The glaciers, which float on ocean water in places, are melting, and when they melt far enough the ice sheets will begin a catastrophic flow towards the oceans that will have the effect of raising the ocean's levels. The ice in Antarctic ice sheets is sufficient to raise the ocean's levels by 200 feet.

Ocean levels are already rising. During the 20th century, Earth's ocean levels rose by 6 inches which was enough to reduce east coast beaches by 50 feet. Loss of shoreline contributes to the devastation caused by storm surges accompanying hurricanes. Satellite data allow geoscientists to track the erosion of glaciers over decades. The Pine Island glacier is one of the largest in Antarctica. It is located on the western end of the continent where it is more exposed to warming ocean waters than other parts of the ice sheet. Erosion of the glaciers is ongoing and increasing. In 2017 a 2,300 square mile iceberg calved from Pine Island. In 2018 a total of 116 square miles of ice broke off of Pine Island in the last weeks of October, the largest being 87 square miles which is four times the land area of Manhattan. Glacial calving of icebergs is caused by emissions-driven global warming. In addition to warming from above, warm upwelling water is eroding glaciers from below increasing their melting rate. Icebergs have historically broken off from the margins of the glacier, but due to melting from below, glacial ice shelves are now breaking up from the center indicating how much more rapidly melting is occurring.

Cry havoc and let slip the dogs of war

The root cause of all the catastrophes besetting the world today is humanity's inability to resolve its fear of death. Fear of death is hidden deep in the unconscious. The vulnerable newborn is motivated by an innate drive to survive. Conscious awareness has not yet formed in the newborn, and its primary process thinking is most evident. Hunger and thirst threaten the baby's survival, and it cries and screams if it is not fed. These motivations are set in the biology, the physiological processes of the newborn, and depending upon how they are managed by mother and father, they become the basis for later behavior. Isolation, separation from the protective presence of parents, is particularly threatening to infants because death was the likely outcome for babies left alone in the prehistoric times in which we evolved, and if not soothed in infancy, the adult that infant becomes will be driven by a fear of death of which it is unconscious.

Every area examined in discussing the Contagion—greed, self-absorption, property, money, misogyny, entitlement, addiction to accumulating—have all been traced back to fear of death. Nowhere is this clearer than in human being's addiction to war, and although war is pervasive in almost all so-called developed societies, it reaches its apotheosis in America. War and killing in war externalizes human fear of death whether it is the bombing of a village in Afghanistan by USAF jets or the suicide bombing of a marketplace in Iraq by ISIS. The equation is brutal and stark: "I kill you so I am not dead." A corollary is: "I make you fear for your life so I am less fearful for my own."

The analysis in this section draws on a long article Medea Benjamin and Nicolas J.S. Davies, writing for Code Pink, published in September, 2018 entitled "War Profiteers: The U.S. War Machine and the Arming of Repressive Regimes."

Corporations and bureaucracies assuage human death anxiety because humans who work for them partake of the corporations' immortality. Five of the largest corporations in the world comprise Big Defense: Lockheed Martin, Boeing, Raytheon, Northrop Grumman, and General Dynamics. The U.S. Military constitutes a gigantic bureaucracy. The corporations and the bureaucracy exist to defend Americans against their death anxiety. The amount of its national wealth and productivity that America pours into defense accurately measures the enormity of Americans' death anxiety. On August 1, 2018 the United Senate gave its approval to a $716 billion defense budget that accounted for 20 percent of the $4.4 trillion U.S. budget for 2019. The money will be used to maintain 800 military bases in 80 countries world-wide. The U.S. government through its military spreads chaos and death to all of these 80 countries that is only different in the enormity of its scale from the chaos and death the U.S. military acted out on the Indigenous people it was conquering in the seventeenth and eighteenth centuries. A massive wall of denial blocks the nation's awareness of what is being done in the nation's name and what is being done with the money the nation could alternatively invest in education, health care, income equality, and infrastructure. The nation accepts the euphemisms the military-industrial-political class uses to hide the true reality of what is being done with the nation's wealth. Citizens accept that the world needs the police force America provides. American citizens are willingly if unconsciously investing the nation's wealth—one trillion dollars in the Iraq war alone according to Nobel economist Joseph Stiglitz—in the murderous defense against death anxiety. Citizens are shielded from recognizing the true purpose of the military-industrial-political complex: to externalize the fear of death Americans feel, to murder human beings elsewhere to give Americans a sense of power over their own mortality.

Benjamin and Davies write that the military-industrial complex Eisenhower warned the nation about in his speech upon leaving office in 1961 is the cause of the plundering of our nation's treasury and the carnage the nation wreaks upon the world. This is like saying the syringe is the cause of the addiction's heroin addiction. The military-industrial-political complex is the means by which the true cause manifests itself. The true cause is America's fear of death, the cumulative death anxiety of America's 325 million citizens and most of the other 7. 2 billion humans on the planet. In the collectivity of the world's nations, America and its clients and agents have taken the role of murdering and terrorizing poor, impoverished, defenseless people world-wide to ease the death anxiety of its own citizens and the citizens of all the nations that acquiesce in America's actions. All humans employ verbal defenses to justify and explain our murderous military actions, but in fact it is the unconscious chthonic motivations that cause us to murder and terrorize others as a defense against feeling terror ourselves.

Davies and Benjamin estimated the enormity of the murder of fellow human beings by America and its allies since the attack on the World Trade Center that resulted in the deaths of 3,000 American citizens. The destruction of the World Trade Center powerfully elicited Americans' fear of death, and the nation set out on a killing spree in an effort to ease its own death anxiety. The attack on Iraq claimed 2.4 million human beings and the attacks on Afghanistan and Pakistan took 1.2 million more human lives. The attacks on Libya resulted in the deaths of 30,000 human beings when the findings of several studies are combined. The numbers of deaths in other recent conflicts for which the U.S. or its clients are responsible are not included in these totals. U.S. supplied weapons have brought death to Angola, Bosnia, Democratic Republic of Congo, Guatemala, Honduras, Kosovo, Rwanda, Sudan, and Uganda. When the U.S. attacks a country, the fear of death buried in every citizen's unconscious in the attacked nation is released, and they begin externalizing their fear of death, murdering each other in a frenzy of killing driven by their own death anxiety. Once unleashed humans die from many causes besides bombs, missiles, phosphorous, and snipers: disease, starvation, and exposure.

Benjamin and Davies report how the U.S. government in concert with the Big Defense Corporations have facilitated the governments of Saudi Arabia in bringing death and destruction to Yemen, of Israel in slaughtering Palestinians, and of Egypt in massacring 2,600 civilians peacefully protesting in Cairo's Rahaa Square. The U.S. government pays for the weapons and the Contagion-driven corporations of Big Defense greedily siphon off profit for their shareholders and executives. Propaganda produced by the U.S. government hides the reality of the death the weapons produce in a massive denial strategy that keeps the reality buried in the nation's unconscious.

Killing motivated by fear of death has reached a high point in modern times perhaps due to the lethality of contemporary weapons, but it is a process that has been ongoing since pre-historical times. The earliest reported war occurred in the Bronze Age (13th and 14th centuries B.C.) when Troy and the Greeks fought. Herodotus estimated that the Trojan war claimed 30,000 lives. All of the citizens of Troy perished, most from diseases: dysentery, malaria, hunger, exposure, and various plagues. The Peloponnesian war between Athens and Sparta took place between 431 B.C. and 404 B.C. and was reported by Thucydides. In that period the dead were not counted and data were not collected. Tens of thousands apparently died as a direct result of fighting and the ancillary results of disease. Rome fought several wars, most notably the Punic war with Carthage between 246 B.C. and 146 B.C. that claimed an estimated 600,000 men's lives. Warfare consumed ancient Rome as fully as it currently consumes America, and in fact has consumed almost every nation in the period from then to now. Between 1618 and 1648 Protestants and Catholics fought the Thirty Years war in central Europe. Fatalities numbered 8 million from both military battles, violence, famine, and plague. Ushering in the twentieth century, the process of killing to assuage death anxiety became global. The First World War resulted in 20 million casualties about half of them soldiers and half civilians. The killing reached a high point in the Second World War which is believed to have killed as many as 80 million people, about 3 percent of the world's population at that time.

The enormity of these numbers supports the assertion that human beings murder each other because they fear death themselves. Accepting this assertion—even being able to consider it—will require a deep level of awareness, one that penetrates below the ordinary awareness of the historical narrative which casts wars in terms of conquest, territorial expansion, protecting the homeland, domination, and fear of minorities. Every one of the conflicts mentioned had a justification. They all do today. The conscious mind provides an excuse for what the unconscious is already executing. *Requiem* asks readers to explore more deeply. Here is a brief experiment to help you examine your own unconscious motivations.

Imagine that your urban city of 200,000 is considering its budget for law enforcement. Which of the following will you choose right now in a referendum on the proposal. Please circle your answer.

Decrease the budget   No change   Increase the budget

Now, read the following paragraph.

Death begins the moment the heart stops beating. Without a beating heart, the blood begins to pool wherever it is in arteries and veins. The body begins to change color, becoming purple where blood has pooled and ashen white where blood has drained out. Bodies lose their temperature and cool down to the ambient temperature. About four hours after the heart stops beating, the muscles contract and the body stiffens (rigor mortis). The dead muscles can contract leading to a twitch. The facial muscles smooth out. The body loses moisture and the skin dries out making it appear that hair and nails are growing. The sphincters in the intestinal tract relax and the bowels empty. As bacteria multiply and begin to decompose the body, putrescine is released, an extremely noxious chemical. In the wild flies, vultures, and coyotes tear the body apart and maggots go about consuming the soft tissues. The internal organs liquefy. The skin becomes dry and leathery.

First, what feelings and thoughts came up for you as and after you read the paragraph. Write them down.

Second, imagine once again that your urban city of 200,000 is considering its budget for law enforcement. Which of the following will you choose in a referendum on the proposal now that you've read the paragraph. Please circle your answer.

Decrease the budget   No change   Increase the budget

What did you answer after you read the paragraph on what happens to the body after death? Was it the same, or did you budget more money for law enforcement? Or less? The paragraph induces a state called "mortality salience" in which unconscious thoughts, feelings, and attitudes about death are triggered because readers have become more aware that death is inevitable. The effect of awareness of death on a decision relating to protection is measured by the amount budgeted for law enforcement before and after reading the paragraph.

This chapter began with an image of the doctors presenting their evidence that the patient, the world, is moribund and that there are no heroic efforts that could be undertaken that would save the world. The doctors have explained that the tipping points have all been reached. That the world is on a course for a seven degree Celsius temperature increase even if we begin to reduce CO_2 immediately. We have already overpopulated the planet. That cannot be undone. The chapter offered a model for why there are no workable heroic efforts. The basis of the model is the power of unconsciously motivated death anxiety that is inextricably tangled in every aspect of the Contagion that drives our society. The chapter argued that Contagion is a population effect and that a majority of people will have to change if the society is to change. At this point, the family of the patient would ask the doctors, "What is your advice? What should we do?" If we were dealing with an individual dying human being the doctor would ask, "Has the patient signed an Advance Directive?" An advance directive expresses a patient's request to forego heroic measures. Earth and humanity have signed no such Advance Directive. In the absence of an Advance Directive stating the patient's will, the doctors advise that the patient be transferred to hospice where its physical, mental, emotional, and spiritual needs can be attended to as the patient enjoins the death process. Chapter Five focuses on the mental, emotional, and spiritual components of hospice for the world.

Chapter Five

Requiem

Chapter Four asked readers to consider that the progression of Contagion had reached a point where no heroic measures could save the world. It argued that the tipping points have all been passed, that the world's death process has acquired a momentum that cannot be stopped or reversed. The ravenous Contagion of greed and power and money and the denial strategies are too great. On the day civilization collapses the stock brokers will still be on the trading floor trying to make more trades, the military will still be attacking an enemy, a politician will still be trying to pass legislation, a commuter will once again be stuck in traffic. The situation is analogous to when a human's cancer reaches the pancreas. At that point there is no remedy. For our world, the corporations and bureaucracies are like the cancer, and they will not stop. The politicians and military will not stop. People are entranced, and they cannot stop what they are doing. They are stuck in the insanity of doing the same thing over and over and expecting a different result. The insanity has not been raised to conscious awareness. If it had, we would have realized in the middle of the twentieth century that we had been infected with the Contagion and we would have changed our behavior. Although we are not consciously aware of the Contagion or our approaching death, at an unconscious level people know the world is dying. As a global civilization we have entered the grief process. *Requiem* hopes to raise our grief process to awareness.

Stages of grief

Elizabeth Kubler-Ross studied death and dying. She famously identified five stages in the process: denial, anger, bargaining, depression, and acceptance. Applying this model to the Contagion-infected world, our civilization is stuck in the denial stage. Congress, the President, the corporations, and the government bureaucracy all act as if there is no problem. Trump has called climate change and global warming a hoax. Senator Inhofe's snowball stunt exemplifies the denial phase. The celebrations by shipping companies of the melting of the Arctic ice without the least care that global weather patterns are changing because the Arctic ice melting, these too demonstrate the denial stage. The anger stage can easily be uncovered when the climate change denier's thoughts and beliefs are challenged. Very quickly the denier's voice will be raised, his body posture will stiffen, he will jab at the air, and he will begin to drop F-bombs in his speech. Some parts of society appear to have shifted from denial into the bargaining stage. California, for example, has committed to 100 percent renewable energy by 2045. That action is significant. It is also too little and too late. Change in California is positive, but California has a population of 35 million. The world's survival requires a change in thought, feeling and behavior of the rest of the planet's 7.5 billion people. Moreover, California's commitment does not address the thoughts and emotions and the deep unconscious motivations that drive climate change.

Most climate scientists fear that Earth's temperature will probably hit 7.2 degrees Fahrenheit (4 degrees Celsius) above the baseline by the end of this century. The current fear is that the planet has passed a tipping point where methane held in the permafrost and in semisolid form at the bottom of the ocean will explode to the surface and the planet could warm by 6 degrees Celsius (10.8 degrees Fahrenheit). These temperature changes will make Earth uninhabitable for humans and for most other species. Presently only three nations on the planet obtain 100 percent of their energy from renewable resources: Albania, Paraguay, and Iceland. Incompletely effective measures identify the bargaining stage. It is like the addict saying, "I know I need help, so I'll go to meetings and I'll learn to use in moderation." The process of global warming has begun and cannot be changed with triage.

The depression stage of grief follows denial, anger, and bargaining. Depression ensues when people become aware that attempts that characterize the bargaining stage of treating symptoms instead of causes have not worked. In the depression stage, people withdraw into grief, withdraw from social contacts, isolate themselves, and often weep. Anguish is the salient emotion of the depression stage. A person in depression feels helpless, hopeless, and overwhelmed. In the depression stage of the grief process for the death of our world, a person feels helpless to reverse the course of coral reef die-off, hopeless to stop people in America and Asia from contaminating the ocean with plastic, powerless to stop Big Energy and its political accomplices whether in California, Texas, Australia, or Canada from fracking and pumping the poisonous waste into the water supply. Watching the news coverage of increasingly intense hurricanes like Florence and Michael, and Asia's Mangkhut in 2018, the depressed stage person's emotions are blunted. This person is less likely than before to join a protest, or to sign yet another online petition or to write or call their Congressperson.

The final stage of grief identified by Kubler-Ross is acceptance. The grieving person manages to construct a new meaning-making model, one that confers acceptance of the new reality. People all had meaning-making models that guided their lives and kept the chaos at bay in the time before the Contagion worsened to the point that the world began to die. People's meaning-making models comprised many of the components of the Contagion. They bought and sold real estate for example and that brought meaning. Or insurance. They became politicians or bankers. They never saw themselves as greedy or entitled or self-absorbed. They never considered the consequences of their long commutes or carbon footprint. Not until it was too late. When the meaning-making models collapse, people sink into the depression stage of grief. Without a model to organize their lives, some people slip into psychosis and commit insane acts like the mass shootings that are becoming more and more common in America and elsewhere. If people can sit in the depression stage they may come to a new meaning-making model, one that organizes their lives in a healthier way. Acceptance does not say, "It's okay that the world is dying." Rather acceptance says, "The world is dying and I will die along with it, but I will be able to tolerate that." Anguish subsides as one's emotions begin to stabilize. One enters reality. One comes to terms with the reality that the world is dying and that one will die with it. One accepts the inevitability of death. Acceptance is a time of adjustment and readjustment. There are good days, there are bad days, and then there are good days again. Acceptance does not mean you will never have another bad day where you are uncontrollably sad, but the good days outnumber the bad days. In the stage of acceptance you may lift from your fog, you start to engage with friends again, you might even make new friends. During this time you understand that your world is dying and can never be replaced, but you move, grow, and evolve in the new reality. The remainder of this chapter and all of the book's final chapter intend to help you create a new meaning-making model for yourself.

Ritual

Our civilization has few effective models, few effective protocols for moving through the stages of grief from denial and anger through bargaining and depression into acceptance. We are trapped in the insanity of doing the same thing over and over and expecting different results. Ritual is the means by which individuals and societies transform themselves. As individuals we transform ourselves from single to married through the ritual of the wedding. For women the wedding ritual includes the bridal shower. For men it includes the stag party. We transform ourselves into parents with a baby shower. We transform ourselves when a relative dies through the ritual of the wake or funeral. When a baby is born, the ritual of baptism converts the infant into a member of the church. If the baby is a male, in some cultures, circumcision transforms the infant into a member of the society of men. The men's movement called attention to the ritual of fraternity hazing or military boot camp designed to transform what Jung called "boy psychology" into adult male psychology.

As a society we transform ourselves through rituals as well. We humans have changed ourselves through ritual but with little awareness of what we've done. Through the ritual of the Nineteenth Amendment, America raised to awareness the unfairness of denying the vote to women. Through the ritual of the Emancipation Proclamation and the Fourteenth Amendment the maltreatment of African Americans was raised to awareness, and they supposedly achieved equal protection under the Constitution. Unfortunately the great public ritual of the Fourteenth and Nineteenth Amendments has done little to change society's attitudes and behavior towards African Americans and women. These are what the men's movement called "false rituals," which like hazing into a fraternity or boot camp do little to transform the old psychology. These examples illustrate the all-but-insurmountable challenge America faces in trying to reverse the Contagion attitudes that have led to climate change.

Having no truly transformative rituals of its own, America can turn to the example provided by Indigenous people of North America who faced a cultural cataclysm in the seventeenth to nineteenth centuries that was every bit as dire as what America and the world are facing at present. Near the end of the nineteenth century, the Native People living in what is now called America knew their culture was finished. They had lived in North America—they called it Turtle Island—for 40,000 years or more and had created a stable, vibrant society that thrived in a sometimes hostile environment. Their way of life was interactive with the environment, reverential, and respectful. They lived in harmony with Great Mystery and worshipped Our Mother the Earth and Our Father the Sky. Rituals and ceremonies organized their lives. The environment provided for them, and they gave back to the environment. Warfare was not unknown, but they did not kill each other gratuitously, they did not attempt to destroy each other's people, tribes, and way of life.

Arriving Europeans brought an entirely different way of life. From the moment Columbus first landed on Hispaniola, he and his men began to kill the native people they encountered. The Europeans were infected with Contagion. The Europeans killed without reason. The Arawak people of Hispaniola did not make war on the Europeans, yet the invaders killed them. The Europeans did not respect the life, the humanness, the sacredness of the people they encountered. The Europeans brought Contagion with them, a disease of behavior, emotion, and spirit. The Native People named this contagion "Wetiko." The Contagion is a mindset—it exists today—of money, property, self-absorption, dominance, entitlement, greed, and violence. Wetiko is toxic individuality. The Contagion whispered in the European's ear that reality serves the strongest, not the most cooperative. The Europeans certainly saw how the Arawak lived communally, helped each other, and lived a peaceful, spiritual, abundant life. The Europeans hated what they saw. The Contagion is anti-spiritual. Contagion hates spirituality. Contagion does not accept mutuality. It only recognizes force and domination. It is sneaky, duplicitous, and mendacious. The Contagion the Native People called Wetiko is opposed to every aspect of traditional Native ways of life.

Indigenous people had no defense against the Contagion, just as they had no defense against the other diseases the Europeans brought: small pox and measles. Within two centuries, the population of Indigenous people was decimated. Their culture was purposefully destroyed, because Contagion told the Europeans that native culture was inferior. Native People were prevented from speaking their languages, they were denied their rituals and ceremonies, their drums were taken from them, their *metates* were broken, their lands were seized, their children were removed to indoctrination schools, and they were forcibly relocated again and again to uninhabitable places.

Wovoka, a medicine man of the Paiute people, who created the Ghost Dance, understood that Native Peoples and their way of life were dead. He understood that greed and money and the concept of property had contaminated Turtle Island. The Contagion was a Black Hole that could not be resisted. Nothing could escape the psychic pull of dissociation, the dissociation between Europeans and the natural world, between Europeans and the rest of humanity. The psychic Black Hole sucked in the Native Ways. Wovoka and the Indigenous People still alive near the end of the nineteenth century knew their old ways were gone for good. The culture was destroyed, their way of life ruined beyond repair. Wovoka received a vision of a ritual to celebrate the culture that was now disappearing, that was now being supplanted by the Contagion-driven culture of the Europeans. Wovoka called the ritual Ghost Dance. It was a ritual that embodied the acceptance stage of the grief process. It was one last ritual to be carried out before the Indigenous people and their culture died off forever.

The interpretation of Ghost Dance offered here is not the interpretation that European historians and anthropologists gave to the ritual. The European historians interpreted the Ghost Dance ritual from the perspective of the conquerors who believed they were superior, more deserving, more chosen of their God whom they believed gave them dominion over the lands they were taking and over the people they were conquering. The European historians interpreted the Contagion from *inside* the Contagion. European historians believed Ghost Dance was the last frantic effort of a defeated people to call down power from their "god" to defeat the invading Europeans. The Europeans told themselves the

Indians believed the dance would empower the Indigenous people so that bullets would pass right through their bodies without injury. Europeans misunderstood and misinterpreted Ghost Dance.

The word "ghost" is being used here to refer to awareness of a being who was once alive and aware and is no longer. Any once living thing can become ghost once it has passed over. American society is populated with ghosts. Although no person alive today actually knew George Washington, American culture keeps him in awareness as a ghost. Jesus is a ghost held in awareness by Christians the world over. So is the Buddha. So is Muhammed. John F. Kennedy qualifies for ghost status. The more recently deceased Prince, David Bowie, and Aretha Franklin exemplify famous people in the process of attaining ghost status. If your parents are deceased you may hold them in awareness as ghosts. Any once living thing can be a ghost. Trees can be ghosts after they have been killed. Northern California is densely populated by ghosts of redwood trees. Our society carves out small islands of untouched redwoods to help us hold in awareness the once vibrant, expansive nation of redwood trees that now have become ghosts. Ghost Dance was a ceremony of awareness by the still-living of the lives, culture, beliefs, and existence of those beings who are no longer alive and aware. The name "Ghost Dance" implies death with awareness.

Now the culture created by the descendants of the first Europeans to come to North America is collapsing. The culture was infected by Contagion, and it is proving to be unsustainable. The unsustainable culture brought by Europeans and infected by Contagion has spread around the world and is killing the world. The culture spawned three new powerful agencies, agencies that are far more vicious, violent, duplicitous, mendacious, and anti-spiritual than Columbus and the European settlers who first invaded the Americas. These agencies are the corporation, its relative the bureaucracy, and the machine. These three agencies have reached the pinnacle of their development in America, and from here they have spread around the globe spreading the Contagion as they go. Just as the small pox and measles brought by the Europeans infected and decimated the Indigenous populations, so the corporation and its twin the bureaucracy together with the machine in all its myriad manifestations are poisoning and contaminating and destroying the environment human beings require in order to survive. The planet is overpopulated. Resources are depleted. Fisheries are collapsing. The planet is warming. Vast tracts of plastic float in the middle of both the Atlantic and Pacific, break up, sink to the ocean floor, and contaminate the fish, whales, and birds. As species go extinct, Homo sapiens itself is threatened with mass die-off if not outright extinction. Few humans alive today could survive without the culture that is both killing them and that provides for them.

We twenty-first century descendants of the Europeans who brought the Contagion to North America when they invaded need our own Ghost Dance. As we clear the Contagion from our eyes and minds we become aware of what we are losing as the world dies. Requiem is our Ghost Dance. We twenty-first century citizens have a choice. We can choose to die in Contagion, blind and unaware, denying and angry that our comfortable world is collapsing and vainly trying to make some compromise, some bargain that will allow us to continue to live in Contagion a little longer. Or we can choose death with awareness, celebrating that which is dying, taking responsibility for our part in destroying this beautiful world. Requiem comprises all five stages of the grief process. It guides us out of denial, honors our anger at ourselves for what we have done, shows us the futility of bargaining, and supports us in our wretched anguish. Requiem leads us into acceptance. Requiem provides an opportunity to die in Right Relations. Requiem will not reverse global warming. Requiem will provide a new meaning making model and an opportunity for death with dignity and acceptance.

Requiem is a ritual that will help us accept the shame we humans feel but deny, our shame about what we have done to the world. Allow yourself to think for a moment about the plastic contamination we humans have spread throughout the oceans. A simple internet search will show you pictures of the gigantic gyres of plastic in the Atlantic and Pacific and will show you underwater images of plastic in the waters off the coast of Bali. Notice your emotions and sensations as you view images of birds, fish, and whales killed by plastic they have swallowed. If you can suspend your denial for a moment you may be aware of a queasiness in your stomach that is the signal you're feeling disgust. You may be aware of a collapsing in your chest that tells you you're feeling shame. Our shame is so painful that most cannot acknowledge it, cannot feel it, and so we deny it. Like a drug addict who denies that his using is causing problems, we humans deny that our overpopulation of the planet, overfishing the oceans, deforestation of the land, pollution, carbon emissions, and global warming are problems, and we continue to engage in those behaviors. We say if the corporations of Big Energy would stop fracking and burning coal the skies would clear and global warming would cease, and we avoid taking responsibility for our habits, our unwillingness to change our lifestyles that result in the polluting caused by our carbon footprint.

One of the great teachings of 12-Step is that only individuals can recover. If every alcoholic or addict waited for all addicts and alcoholics to recover, nobody would recover. Applying that lesson to our species and our civilization, we realize that our civilization may never heal its infection with Contagion. Our civilization may never cease its dependence on oil or its use of plastics or change its money economy. The Sixth Extinction is so well underway that even if we did change our behaviors, the consequences are set in motion.

What we learn from 12-Step is that recovery from Contagion is an individual effort, and that benefits will accrue to the recovering person just because that person has ceased the addictive behaviors. Find a new way of life! That's the promise of 12-Step. Recovery from Contagion is a spiritual awakening, a change of personality. Discover a way of life free from acquisitiveness, free from greed, free from competition. You will invent this new life for yourself as you go along.

Western civilization has no history of ritual practice, no spiritual practice to fall back on as we struggle to resolve Contagion. Religion is a part of the problem of Contagion. Religion is a ritual that has not and will not help us to resolve the problem of Contagion. Requiem on the other hand, ritually celebrates the death of our own cultural "old ways." Chapter Six will propose new ways of being that we can replace the old ways with. Here we will begin the letting go of the habits, the addictions of Contagion. The Requiem also celebrates and honors the good things, the beautiful things we have lost due to our Contagion. Through the ritual of letting go that is Requiem we become aware of who we have been and how we created and became the Contagion. In 12-Step this work comprises Steps Four and Five. Step Four says "We made a searching and thorough moral inventory." Step Five says we admitted the exact nature of the ways in which we violated our moral inventory. This part of the Requiem is a writing exercise. Recovery from Contagion asks you to write your experience of each of the elements of Contagion that have infected your own life. Through the ritual of examining your life you will raise to awareness the unconscious schemas that have driven you as a participant in this civilization, and you can then assess how well they are serving you.

This exercise will not be easy to accomplish. You are dealing with your most deeply held unconscious schemas, and your identity, which is to say the survival of your self-concept, will be challenged by the work. Remind yourself that you are raising these beliefs to awareness. Renew your awareness that you have undertaken to examine these beliefs so that you may celebrate them and then allow them to die.

Death of limitless growth and expansion

How has the belief in limitless expansion affected your life? For example, do you own stocks in a manufacturing corporation? Ask yourself how the income you receive from your investment depends upon the company's attempts to grow without limit. In the very early days of automobile manufacturing, apparently limitless expansion existed because nobody had cars because they didn't exist. When cars were invented, everyone wanted one. The market for automobiles was untapped. However, once the market was saturated, manufacturers had to invent new ways to grow without limit. One way was to produce ever newer models of cars and then to stimulate sales with advertising. There was another way: planned obsolescence. Robert S. McNamara, a young executive at Ford Motor Company, conceived of designing a car so that it would become obsolete after an uneconomically short period of time. McNamara wanted people to have to purchase a new car every 3 to 4 years, and to do this he invented the idea of an expiration date. Equipment was designed to wear out after a few years of use. It had to be replaced. Of course, outside the automobile industry this marvelous invention was not widely advertised. Consumers could not be told how they were being hustled.

Planned obsolescence is the basis of consumerism. It pervades our society. Fashion is another example of planned obsolescence. Fashions are made obsolete by the steady appearance of new fashions. Attitudes are promoted among consumers that tell them they must have the latest design or risk embarrassment and shame. Planned obsolescence and consumerism are Contagion attitudes that result in discarding perfectly good cars, clothes, houses, toys simply because they've become obsolete. Most people do not realize that obsolescence is programmed into the system. Consumerism generates enormous amounts of waste and uses resources that are increasingly scarce. Consider for a moment the single use plastic bag. Consumers purchase a package containing a roll of zip lock plastic bags. They use a bag once and throw it into the trash. Often the plastic bag ends up in the ocean where it contributes to the Texas-sized spinning gyres of plastic rubbish whirling in the Atlantic and Pacific. Filter feeders swallow it, and it clogs their digestive tracts and kills them. As it degrades it ends up in the intestinal tracts of fish.

Now, examine how consumerism expresses itself in your life. And as you do, begin the process of grieving the loss of the latest fashion, the newest automobile, the most recent electronic device, a new TV. Notice the feelings that come up when you imagine not having the newest iPhone or a flat-screen TV.

Death of self-absorption, entitlement, and exceptionalism

Self-absorption is a mostly unconscious schema that causes people to think solely of themselves. Entitlement is a behavior that expresses self-absorption. Self-absorption and entitlement are key elements of narcissism. On a national scale, entitlement becomes exceptionalism. Exceptionalism can be national, as in "American exceptionalism" the idea that America is the most favored nation, that all nations should follow America's example and accept America's form of government. Closely related is the idea that capitalism is exceptional and that America is justified in imposing capitalism and defeating socialism or any different belief system. The three attitudes of self-absorption, entitlement, and exceptionalism also motivate religious fervor especially when it becomes proselytizing. Christians often become so self-absorbed that they cannot accept that any person could believe otherwise and that all people must accept Christ as their personal savior. Muslims often believe Islam is the one true religion and that all people must accept Muhammad. Underneath these attitudes and behaviors lurks death anxiety. People infected by religious Contagion cannot accept the existence of different beliefs. Millions were slaughtered in wars fought over conflicting belief systems. Three crusades killed Middle Easterners who were judged to be infidels because they were not Christians. A fourth crusade killed tens of thousands of Europeans living in the south of France because they believed Jesus and Mary were married, and this threatened the mortality of the Catholics who believed the church's dogma.

Examine your own attitudes of self-absorption, entitlement, and exceptionalism. How do these attitudes appear in your relationships? In a marriage entitlement can manifest as believing a wife should obey her husband and do as she is told. Self-absorption denies a wife the freedom to have her own life, to make her own choices of friends, to go where she feels like going without first obtaining her husband's permission. In family life and child-rearing, self-absorption says children should be seen and not heard. Self-absorption denies children the freedom to explore and pursue their own identity, to find themselves. Self-absorption tells children, "It was good enough for me so it's good enough for you." Self-absorption says, "I was beaten by my father, so I will beat you." Self-absorption punishes children rather than imposing consequences. The self-absorbed parent says, "I'm beating you for your own good." The self-absorbed parent attacks, criticizes, mocks, and belittles.

In regard to the environment and the world, self-absorption expresses in many ways. Ask yourself if you believe wolves should be hunted to extinction. What about coyotes? Do you believe bears should be exterminated? Do you believe hunters should be free to kill hibernating bears and their cubs in their dens? Should people be free to kill rhinoceros for their horns? Elephants for their tusks? Examine your attitudes of self-absorption, entitlement, and exceptionalism. As you uncover your own experiences of self-absorption and entitlement, begin a process of grieving the loss of these attitudes as you release them.

In American culture gun ownership exemplifies defense against death anxiety as much as or more than any other Contagion-driven mass behavior. As has been noted, U.S. civilians own 393 million firearms, which works out to 1.2 firearms per citizen. The literature published by the N.R.A. demonstrates the self-absorption and entitlement of gun owners. Gun owners' reactions when firearms ownership is challenged illustrate well the grief process which indicates the extent to which guns and identity are intertwined. The first stage is denial, and gun owners will deny that guns are a problem or even that guns cause gun deaths, beliefs that are directly contradicted by data. The anger stage of grief seethes just below the surface of the gun owner. There is very little bargaining with an owner of firearms. The N.R.A. even opposes limitations on background checks for weapons sales. For the purposes of this work, try to avoid thinking that the exercise is asking you to relinquish your weapons and instead focus on the thoughts and attitudes that lie beneath your gun ownership. What images come up for you when you think about needing to use your firearm? If you own an assault rifle, what images are associated in your mind with your AR-15? Write them down. Try to just observe rather than engaging in defensive ideas. How is your identity, your personality merged with gun ownership? Do you share a bond with your male friends through gun ownership? Do you and your sons—or daughters—bond through the act of shooting and owning guns? Write about your sense of entitlement to own a weapon. Look at your attitudes that surface when you are asked to consider that children are much more likely to shoot each other or themselves with guns or that men are much more likely to shoot their girlfriends or wives or themselves with guns than they are to shoot an intruder. Take a moment and try and imagine who you would have to be to let go of your attitudes toward guns. Attempt to see how the Contagion expresses itself in the area of guns and gun ownership and what the issue of gun ownership reveals about the Contagion that is destroying our world.

Death of money, property, corporations, and bureaucracies

Denial of death unites modalities of money, property, corporations, and bureaucracies. Money and the greedy accumulation of money feel like a defense against death anxiety. Property is a means for accruing money. Corporations and bureaucracies are organized social structures that facilitate, manage, and organize the activities of accumulating money. The Contagion that Columbus and then the pilgrims brought to America was driven by competition for money. Columbus sailed to the New World to find gold, and he murdered Arawak natives when they could not produce the gold he believed they had. The European culture the pilgrims brought was entirely the opposite of the mutuality the Indigenous people lived. Contagion displaced Indigenous civilization. What are your beliefs about money? What did you learn from your parents about money? How does the pursuit of money and property affect your life and the lives of your children? What feelings come up for you if you imagine for a moment that the company you work for closed down? Suppose for a moment that you have money invested in the stock market and examine what feelings arise if the stock market were to crash. Do you feel fear? Terror? What images are coming up along with the emotions? This is death anxiety. Notice how money and its pursuit have insidiously taken over your life.

Now, to get a sense of the extent of the Contagion allow yourself to visualize the entire world driven by the same fear you just felt, overwhelmed by the enormity of the concept of having no income. From the impoverished people sifting through trash heaps in Asia to the hedge fund managers of

Wall Street, every single person is driven by terror that without money they will die. If you are able to do this, then realize how this monstrous system of money and its pursuit has resulted in the destruction of the world. Pollution, climate change and global warming, glaciers melting, oceans rising, hurricanes, species extinction are all outcomes of Contagion and are all directly related to defenses against death anxiety. Now, imagine for a moment what would your life be like if you were not absorbed in the pursuit of money. How would you feed and shelter yourself and your family? What would you do if you got sick? Who would care for you in your old age? What would you do with your time? How would you entertain yourself? What would you do for fun? What would a society look like that could handle these challenges? What would you have to change and what would everyone else have to change to survive in such a society? What attitudes would people have to have to make a society like this work? As you examine this side of the equation you may begin to realize the challenge of inventing a healthy and sustainable way of living, one that would not result in the death of the planet. Does this feel possible, like it could even happen, like people could change sufficiently to make it work? Write about your relationship with money and property and then consider letting go of money and your obsession with it. Allow yourself to grieve the loss of the meaning money and property have provided you. Consider a ritual you can do to alter your preoccupation with money. Would you be willing to burn a dollar bill in a ritual of transformation?

When you have grasped the enormity of the world-wide disaster and how unlikely it is to change, you will become ready to perform requiem for our dying world. You will be moving through the stages of grief toward acceptance. The following section comprises a "letting go" exercise. The intent of this exercise is to prepare you for the death of the world. You can perform it now and be at peace as the world's death throes worsen. For some people the end will be sudden as when a Category Five hurricane tears the roof off the house and buries them under piles of roof beams. Or it may be sudden as when a fire consumes their house as it destroys a neighborhood or a city. For some it may last hours or even days as the electrical grid goes down and the air conditioners no longer work and the temperature in the house rises to meet the 130 degree Fahrenheit temperature outside. It may be starvation when there is no food on the shelves at the supermarket and no gas to try and drive somewhere in hopes of finding food. It may be thirst when there is no water coming out of the tap or the water that does come out is so poisoned with chemicals that it cannot be swallowed. The choice facing all of us humans is whether we want to die with awareness or not. Death of the world is inevitable. We can choose how we experience that death.

Honoring losses

Requiem is a ritual. This is the first of a few rituals. A ritual is an activity, and this and succeeding sections use the word "requiem" as a verb as well as a noun. As you read this requiem or say it out loud, allow yourself to experience the emotions that come up and to notice images that surface. Allow yourself to become the requiem. Say the requiem slowly. Try not to just read about the requiem. Because this work is essentially a spiritual exercise, begin by saying this brief invocation out loud:

We enter now sacred space
We enter now sacred time
We open to the way of Requiem
We surrender to the way of Requiem
We believe in the way of Requiem
We trust in the way of Requiem.

We requiem for the end of the old ways. There is no more easy oil. We requiem for the end of easy oil. We requiem for nuclear energy. There can be no more divisive politics. We requiem for the end of sectarianism. We requiem for the death of partisan divide. We requiem for the death of selfishness. We requiem for the death of narcissism. We requiem for the death of entitlement. We requiem for the death of individualism. We requiem for the death of private property. We requiem for misogyny. We requiem for the death of male privilege. There can be no more male privilege. We requiem for the death of selfishness. We requiem for our failure to cooperate and to create a society of mutuality. We requiem for the death of corporations. We requiem for the death of the stock market. We requiem for insurance companies. We requiem for the death of hedge funds. We requiem for the death of private property. We requiem for an economy of continuous unlimited growth. We requiem for the death of consumerism. We requiem for the death of the commons. There is no more commons to be appropriated, to be despoiled, to be used for private gain. We requiem for the death of a throw-away, disposable society. We requiem for an uncaring society. We requiem for the death of a society of ignorance, or unawareness. We requiem for unlimited population. We requiem for big families. We requiem for 7.5 billion people. We requiem for a society of competition. The society of competition is dying. The old ways don't work anymore.

We requiem for the death of gun ownership. We requiem for the identity of the gun owner that is becoming ghost. The guns are becoming ghost. We requiem for the death of the belief that firearms will protect us against the Contagion. We requiem for the death of attitudes of entitlement and male privilege that motivate gun ownership. We requiem for the ghosts of all the 17,207 American children who are killed with guns every year. We requiem for the millions and millions of human beings killed with guns since rifles and guns were invented. We requiem for the American and allied soldiers killed with guns in our time. We requiem for the Syrian men, women, and children killed with guns and have become ghost. We requiem for the Iraqis, Sunnis, Shiites, and Kurds, men, women, and children who have been killed by guns, missiles, bombs, IEDs, and land mines. We requiem for Afghani men, women, and children who have been killed by weapons of all kinds and have become ghosts. We requiem for the Chechnyans, Ukrainians, men, women, and children, who have been killed by guns and become ghosts. We requiem for Nigerians, Sudanese, Moroccans, Egyptians, Libyans, Palestinians, South Africans, for all African people who have been turned into ghosts by guns, missiles, and drones. We requiem for Mexicans, Central Americans, Bolivians, Brazilians, Argentinians, Colombians, and Chileans who have been killed by guns and have become ghosts.

We requiem for the ghosts of our environment. We requiem for the ghost of the California Central Valley Aquifer that is being pumped dry. We requiem for the water that once slaked our thirst. We requiem for the layers of clay and gravel deep under the Central Valley that held the water in the aquifer. The clay and gravel have become ghost, collapsed as the water was pumped out. We requiem for the death of the layers that held the water and will hold water no more. We requiem for the immanent death of the Sacramento Valley Aquifer which will soon be pumped dry as the corporations move north having sucked all the water from the Central Valley Aquifer and left a desiccated waste land behind. We requiem for the Ogallala Aquifer that is being pumped dry and will soon be gone. We requiem for the ghost of lakes and rivers around the world that are dying. We requiem for the death of the Aral Sea in Kazakhstan that is becoming ghost. We requiem for the death of Lake Poopo in Bolivia that has died and become ghost. We requiem for the death of the Colorado River that has been dammed and diverted and pumped so that it no longer flows into the Sea of Cortez. We requiem for Lake Chad that has become ghost as it lost 95 percent of its water. We requiem for Owens Lake in the eastern Sierra Nevadas that was pumped to supply water to Los Angeles and is now become ghost. We requiem for all of the archaeological sites and habitats that were destroyed when Lake Powell was created by damming the Colorado River. We requiem for Lake Mead in Nevada that has lost 60 percent of its water and is almost completely ghost as it is drained for the overpopulated areas of Nevada, California, and Arizona. We requiem for the lakes along the Yangtze River in China that are becoming ghost as they are depleted to serve agriculture to feed the overpopulated nation.

We requiem for Vatnajökull glacier in Iceland that is becoming ghost as the planet warms. We requiem for Pine Island and Thwaites glaciers in Antarctica that are becoming ghost as they melt from above and below. We requiem for the Arctic ice that is melting and fragmenting and becoming ghost due to global warming. We requiem for Alaska's Wolverine and Gulkana glaciers that are rapidly becoming ghost as they melt due to summer warming that is a symptom of the planet's warming. We requiem for the European glaciers of Norway and the alps that are becoming ghost as planetary warming melts them.

Food is becoming ghost, contaminated by corporations' Contagion-driven greed for more profit. Herbicide and pesticide contaminated food poisons children with glyphosate and chlorpyrifos and the children become ghost. We requiem for the death of healthy organic food. We requiem for the death of children whose brains have been damaged by poisonous herbicides and pesticides. We requiem for the people who acquired cancers of the breast or prostate, or blood cancers like leukemia, lymphoma, and multiple myeloma from exposure to herbicides and pesticides. We requiem for the men whose testicles have been poisoned by industrial pollutants and whose sperm have become ghost.

We requiem for consumers poisoned by hormones that produced breast and prostate cancer.

All over the world animals are becoming ghost as the Contagion drives extinction. We requiem for butterflies and bees and all other insects who have become ghost due to spraying with neonicotinoid pesticides. We requiem for corals in reefs around the world as global warming and acidification of the oceans causes the corals to bleach and they become ghost. Also in the oceans of the world, starfish are melting and turning to goo and becoming ghost due to Starfish Wasting Disease.

We requiem for the starfish. We requiem for the amphibians, the frogs, that are becoming ghost because they have been infected with chytridiomycosis a fungal disease caused by habitat loss traceable to human incursion. We requiem for the millions of bats that have become ghost due to a fungus that is a marker of the Sixth Mass Extinction of the Anthropocene era.

We requiem for the Black Rhinocerous that have become ghost as the planet dies. We requiem for the elephants turned to ghost because they were murdered for their ivory. We requiem for the whales, the Right, Sei, Humpback, Blue, Grey, that are becoming ghost as they are hunted, and starved because of habitat loss. We requiem for the Grizzly, Polar, and Kodiak bears that are becoming ghost. We requiem for the wolves and coyotes that have become ghost as they are poisoned and shot.

Honoring the Indigenous people who have become ghost

Focusing solely on the part of North America now occupied by the United States, anthropologists estimate that about 5 million Indigenous people lived here at the time Columbus arrived (range from 2.1 million to 18 million). By 1890, only 228,000 were still alive. What that means is that 4,772,000 human beings became ghost over a period of two hundred years. In the year 2000, Kevin Grover, head of the Bureau of

Indian Affairs, made a formal apology to the Indigenous people for the genocide that the United States inflicted upon them. That was an excellent first step, but as we will discuss in Chapter Six it falls very short of a reconciliation process or an atonement process that would involve all Americans. In the following section, the names of the tribes that became ghost are recited in hopes that the enormity of the crime can be raised to awareness. Try not to just skim through this section. Linger with each name and allow images to surface as you do.

We requiem for the Arawak people of Hispaniola who were murdered outright or who died from disease brought by Europeans. We requiem for the Arawak people who have become ghost.

We requiem for the Pequot and Mohegan people who lived in the Upper Hudson River Valley and who were murdered and who became ghost.

We requiem for the Narragansett people who lived on Narragansett Bay in what is now Rhode Island from the Providence River to the Pawcatuck River and who were murdered or killed by diseases brought by Europeans and became ghost.

We requiem for the Wampanoag people who lived in what is now Massachusetts and Rhode Island and were murdered and died from measles and small pox and who became ghosts and whose culture became ghost with them.

We requiem for the Mohawk people who lived in what is now Albany in New York State. The Mohawk were known by their Indigenous name, Kanienkehaka, and belonged to the Iroquois confederation. The Mohawk were murdered and their land seized, and they and their culture became ghost.

We requiem for the Oneida people who lived in what is now upstate New York and who were a member of the six nation confederation of tribes that included the Seneca, Cayuga, Onondaga, Oneida, Mohawk, and Tuscarora and comprised the Kanonsionni, or League of Clans of the

Iroquois people. The Oneida were murdered and died from disease and were relocated to Wisconsin and Canada and many of them and their culture became ghost.

We requiem for the Iroquois (Haudenosaunee) people who lived in the center of what is now the state of New York. The were murdered and their culture was largely destroyed and they and their culture became ghost. Their lands were stolen from them by Europeans.

We requiem for the Seneca tribe that lived south of Lake Ontario at the western end of the Iroquois nation's land. The Seneca were murdered and died from diseases, and they and their culture became ghost.

We requiem for the Lakota ghosts. The Lakota lived in what is now Wisconsin, Minnesota, North Dakota, and South Dakota. We requiem for the 300 Lakota men, women, and children who were massacred in the snow on December 29, 1890 at Wounded Knee and became ghosts. We requiem for all the Lakota people who were killed with small pox in germ warfare. We requiem for the death of their culture that became ghost as it was destroyed by the United States army.

We requiem for the Zuni, the Hopi, the Apache who were murdered in their homelands in what is now the American southwest and who became ghost

We requiem for the Achomawi, Atsugewi, Miwok, Pomo, Cahuilla,, Chemehuevi, Chimariko, Chumash, Karuk, Yurok, and Paiute who lived in what is now California and Nevada and who were dispossessed and murdered.

We requiem for the Salish, Quinault, Yakima, Palouse, Nez Perce, and Chinook people who lived in what is now Washington and who were dispossessed and made ghost.

We also requiem for the African American people who were enslaved and who died and became ghost. We requiem for the Mandé people of Guinea whose leaders were infected with Contagion and sold their own people into slavery. We requiem for the people of Togo, Ghana, and Benin who died on the slave ships, taken from their families, separated from *their children. We requiem for the Akan people of what is now Ghana and Ivory Coast who were brought to America, sold into slavery, and whipped, and beaten and later lynched and immolated and became ghost. We requiem for the Wolof people of Senegal and the Igbo of Nigeria, and the BaKongo people of what is now the Democratic Republic of Congo and Angola who were taken captive and enslaved.*

We requiem for the modern day descendants of the African people who have been killed in America and who continue to be killed in America and who have become ghost. We requiem for Tamir Rice, for Trayvon Martin, and Sandra Bland, and Sam Hose who was lynched, and Malcolm X, and Emmett Till, and Crispus Attucks, and Martin Luther King, and the thousands of African American people who have been lynched or shot or strangled or knifed or set on fire or dragged to death and who became ghost.

We requiem for the death of critical thinking in America that was murdered by the Contagion and has become ghost.

We requiem for the death of public education that once was valued and that taught arts, poetry, classics, and critical thinking. We requiem for a nation that used critical thinking to form beliefs rather than starting with beliefs and stopping there. We requiem for compassion that was murdered by the Contagion. We requiem for humility, honesty, open-mindedness, and willingness that were murdered by the Contagion and became ghost. We requiem for the death of cooperation and a sharing economy that was murdered by consumerism and competition and selfishness and greed and has become ghost.

And now?

Having begun the requiem process raises to awareness the choices humanity faces. We may choose to return to unconsciousness and to enter the death process striving to maintain the ways of the Contagion that have brought the world to this place. Or, we may choose to go consciously into that good night celebrating the beauty that is passing, the beauty of grizzly bears and butterflies, the innocent grandeur of the pristine wilderness, all that we are realizing too late was what gave authentic meaning to our existence and that we have thoughtlessly destroyed through addiction to the ways of Contagion.

Our requiem encompasses more than the loss of the beauty. Our requiem extends to the loss of the Contagion ways, ways that no longer sustain us, ways that we now realize have poisoned us and our world. Why requiem now? Humanity's situation is not as dire as was the situation of the Indigenous People when Wovoka received the Ghost Dance. Civilization has not collapsed...yet. The indicators of the coming collapse are obvious. It has already begun. It is possible, slightly possible, that humans could wake up. Perhaps reading *Requiem* could rouse us from our Contagion-infected trance. It was slightly possible that European invaders could have been awakened by Wovoka's Ghost Dance, but they were not. They were too entranced in the unconscious schemas, too addicted to the ways of Contagion. There are only two alternatives: do nothing and experience the death of the world with terror and anguish and without awareness or complete the requiem process and experience the Sixth Mass Extinction with awareness and spirituality. Death will come whichever choice humanity makes.

For healing, for recovery, we turn to the 12-Step model. Humanity Rising can structure our process of freeing ourselves from addiction.

Chapter Six

Humanity Rising

Requiem has demonstrated that the external behaviors that manifest Contagion arose from unconscious schemas learned in childhood that express as Contagion. What can we do about the unconscious schemas? How can we change who we are? Chapter Six will introduce Humanity Rising, which is a program for recovery from Contagion. Recovery begins with awareness, which has been one of the primary themes of the book. By defining Contagion we've raised it to awareness. We've exposed the origins of Contagion and how it has manifested in climate change and the other myriad consequences of Contagion. The objective of Chapter Six is to bring to readers' awareness a means for recovering from the Contagion that has clouded our consciousness for eons.

The goal of Humanity Rising is not to reverse global climate change. Humanity Rising operates on the same principle as AA in this regard. AA has always been aware that it cannot end alcoholism; it can only help one alcoholic at a time to recover from alcoholism. Similarly, Humanity Rising is aware that it can only help one human being at a time to recover from Contagion. Humanity Rising cannot end Contagion. Humanity Rising cannot stop climate change or global warming. If sufficient people were to engage the program of Humanity Rising, and if they were to do it immediately, there is a possibility the Contagion-driven destruction of our home world could be halted or at least slowed down. However, as with the alcoholic stumbling in an alleyway, consumed by alcoholism, fiending for another pint of vodka, the most infected members of our society, the ones who can make choices affecting all of us, these are the greed addicts who must find recovery. So far they have shown very little inclination to change their ways. Quoting the American author Upton Sinclair, "It is difficult to get a man to understand something when his salary depends on his not understanding it."

Some people might ask, "What's the point of getting into recovery myself if it's not going to put an end to climate change because some people are still fracking and drilling for oil in the Arctic?" Scientists have clearly stated that tipping points have been passed, that the accumulation of CO_2 presently in the atmosphere means the planet's temperature will rise by at least 2 to 4 °C (8 to 12 °F). The Sixth Extinction is underway. Ask yourself what spiritual condition do you want to be in when the extinction knocks at your door? You can choose to die in Contagion or you can choose to die in recovery. Thinking in terms of the five stages of grief, you can choose to die in denial or anger or bargaining, which means you'll die in terror and hopelessness. Or you can choose go through depression and to die in calm acceptance. Climate change and the Sixth Extinction are forcing us to encounter the reality of our mortality. We cannot force or motivate all Contagion-infected humans into recovery. As individuals however, each of us can enter the recovery process. We have the choice: to die in Contagion or to die in recovery from Contagion. This is true whether humans reverse climate change or not. Every addict will die. Every addict has a choice: to die in addiction, or to die in recovery.

Other people might ask what is the urgency? They say the temperatures where they live are not intolerable. They are like the alcoholic who is determined to keep drinking until something catastrophic happens. Climate change deniers are like the alcoholic who can always point to someone who drinks more than they do and has had more grievous consequences. Global warming deniers say, "At least we don't have it as bad as Australia where it's been near 48 °C (122 °F) for days." So far the consequences of global warming have not become severe enough to motivate people to change their Contagion-driven behavior. Unfortunately the consequences are so dire and are accelerating exponentially that we have no time to wait for people to realize they need to change. By the time the consequences begin to be undeniable, it will be too late. The alcoholic who has put off doing something about his drinking is now in the emergency room being told that his liver has failed and he will be dead shortly.

Every addict, whether a greed addict or a money addict or a property addict, or a cocaine or alcohol addict, every addict faces the same turning point. One direction is doing nothing about the problem of the addiction and continuing to live in the consequences, the unmanageability of the addiction. In this direction, people tell themselves, "I've lived this way for all this time, why should I change now?" They marshal their denial strategies and get another DUI or another arrest for possession. Lose another marriage. If addicted to greed or money, they pay no attention as another drinking water source is poisoned; as the CO_2 in the atmosphere inches up to 411 ppm; as the Arctic ice breaks up completely and the first blue ocean event occurs.

Another direction is to change our own behavior. Working the Steps of Humanity Rising will bring you a new peace of mind. You will be freed from the past and will no longer be condemned to repeat it. You will discover an intimate relationship to the natural world and the beings, animate and inanimate, who inhabit it. You will come to accept the death of the world, and you will be grateful you found a new relationship with the natural world before it did die away.

Humanity Rising is a 12-Step program for recovery from Contagion, a program for raising the unconscious fear-conditioned schemas to awareness and for changing them. The Twelve Steps as manifested in Humanity Rising is a program for recovery from the schemas that are destroying the world and threaten the extinction of human beings and all our relations through their effect on our perceptions, thinking, and behavior. We are all fear conditioned. Through a set of twelve steps the program will help you raise your schemas to awareness. It will help you to relieve yourself of those schemas. The program is not easy but it is simple.

Humanity Rising is more than a recovery program, it is a prescription for life, for living life authentically and mindfully and autonomously. One issue with 12 step is that we cannot know what our lives will be like if we stop our addictive process. What we know about addiction is that it always assures us what will happen next. We will get loaded. Or drunk. It is the same with the self-absorption schemas of Contagion. We always know what will happen next. We'll go to work, we'll use more gas, we'll make our house payments, we'll watch TV, we'll do today what we did yesterday. We'll read about global warming and dismiss it as a left-wing plot or a problem to be addressed tomorrow. Recovery through Humanity Rising promises a new way of living. You will discover a way of living without greed, without the avaricious need to accumulate more. You will discover that your addictive belief that you had power over the world was an illusion. The coming Sixth Extinction is shattering the illusion that we have power over the world. No matter how much money you had, you never had power. Your former desire to extract as much from the earth, to make as many products for sale, to make more and more money, to control people, places, and things will slip away. You will discover that people, places, and things never brought you joy. You will discover a new source of joy: living in oneness with the natural world.

As humanity faces its extinction, it is worthwhile to ask what will be our legacy? For all the poetry and art that human civilization has produced, at the Sixth Extinction what will have been our greatest creation? Undeniably it will be the destruction of the world.

If some human beings should survive the Sixth Extinction, what will they transmit to the generations that follow? Contagion? Or a way of living that is free of Contagion? Given the violent scenarios that are predicted for the Sixth Extinction, it seems likely that any survivors will be motivated by the same greed, the same violence, the same competition that conditioned humans' early emergence 250,000 years ago. That will set the same Contagion-driven process in motion as the one that resulted in our current catastrophe.

A generation of young people is finding its voice. Some of these young people have matured to the point where they are seeking and winning political positions. Their message is the same as the message of Humanity Rising. They recognize the Contagion that is killing their world and their future. Many climate change activists are focused on politics as a way of bringing about change. The record of the last 60 years regarding climate change demonstrates that political change has not worked. An article in the New York Times in 1956 called attention to the process of CO_2 accumulation and the effect on planetary temperature. Climatologists and a legion of scientific organizations have warned about the dangers of greenhouse gases, but although panels met and resolutions were passed, the Contagion-driven forces of greed and money prevailed and the Sixth Extinction began. Politics is part of the problem, not the solution. Humanity Rising offers the generation of concerned young people a solution that addresses the root causes of the Contagion, and it offers a solution: the Twelve Steps.

The program of Humanity Rising consists of the twelve Steps and twelve Traditions. Like all 12 Step programs, Humanity Rising has no opinion on outside issues. This is codified in the Sixth Tradition that states, "A Humanity Rising group ought never endorse, finance, or lend the Humanity Rising name to any related facility or outside enterprise, lest problems of money, property, and prestige divert us from our primary purpose." Following the Sixth Tradition, Humanity Rising does not support candidates for political office. Politicians who may support the goal of recovery from Contagion are not allowed to use Humanity Rising's name in their literature or speeches. The purpose of this tradition is to avoid the infighting and recriminations that often go on in political parties. Were it not for the Sixth Tradition, corporations and their lackeys could advertise that they supported Humanity Rising, but not change their programs. The corporations of Big Energy already advertise their environmental friendliness in a process called "greenwashing" but they continue to drill and frack and spill. We are aware that humans' unconscious schemas have motivated much of Contagion-driven behavior both in the corporations that have promoted and produced climate change and in the political parties that support those corporations. Unfortunately Contagion is ubiquitous, and it can also contaminate the political organizations that oppose drilling and fracking and the extractive processes that contribute to climate change.

Unconscious schemas often motivate people's behavior in political organizations both on the left and on the right. One effect of unconscious schemas being acted out in political organizations that ostensibly support a healthy environment is the disruption and fragmentation and infighting that so often destroys them and that is so easy for agents provocateurs to manipulate. The demise of the Green Party into irrelevancy is a case in point. Governments and corporations infiltrate moles or spies into organizations they view as threats. The moles often destroy an organization by promoting dissension through creating factions arguing over an outside issue. Step Six is a bulwark against these efforts. Twelve Step organizations have never had this problem because they abjure taking positions on outside issues. Humanity Rising can expect attempts to infiltrate it and destroy. Tradition Six is our protection.

In order to avoid the dangers that Tradition Six is designed to circumvent, *Requiem* itself will not discuss or recommend or lend the Humanity Rising name to any political party or "outside enterprise." Movements dedicated to protecting the environment and human beings and all our relations are easy to find using an internet search. If your are motivated by your working the Steps of Humanity Rising you will find these movements on your own.

Step One

Step One: We admitted we were powerless over what happened to us in childhood as a result of which our lives became unmanageable.

Life free from Contagion begins with an honest admission of the power that our early conditioning—both as individuals and as a species—has exerted over us, how our early conditioning became the Contagion, and how powerless we have been over the Contagion and its effects on our world. We admit that because of the Contagion our lives have become unmanageable. Earlier chapters described how the challenging environment in which we evolved left an indelible mark on our brains that tells us we need to compete and to accumulate as much as possible even at the expense of others. Unfortunately we humans were not able to overcome our early conditioning, as a species and as individuals, even though we can provide ourselves with food and shelter in ways that our ancient ancestors could not.

Step One codifies two central concepts of Humanity Rising: powerlessness and unmanageability. In a deviation from standard twelve step, Humanity Rising's Step One introduces a psychological foundation for the steps. AA was founded in 1935, a time when modern psychology was in its early formative years, and psychologists did not yet know the effect of early experience on the development of the brain and the personality. Trauma had not been studied. Humanity Rising was founded in 1992, 57 years after AA and Step One were formulated. During those intervening years, psychologists amassed enormous information about child development. Only in the last 25 years has the field become aware of the profound effects of early experience.

Humanity Rising's first step integrates modern psychology into the 12 Step approach by calling attention to the critical importance of a child's experience in its first years to their development subsequently. The lessons learned through observation during the first four to five years of life form the basis of the Contagion, and these lessons are unconscious and are not easily reviewable, and these are the lessons passed from generation to generation. The Contagion will not be stopped unless the present generation begins recovery and teaches the next generation the skills of living free of Contagion.

Powerlessness

Step work is writing work. It is critical to the success of working the Steps that you think deeply about powerlessness and unmanageability. What follows is a list of topics that have been discussed in the preceding five chapters. Write your personal experience of each topic. Allow yourself to feel into each of the subjects and notice your feelings. If you feel resistance or discomfort with the exercise, notice how that is powerlessness too. Write everything down and discuss it with your sponsor.

Step One asks us to examine how we were powerless over what happened in our personal childhood and then to relate that to what happened in the childhood of our species.

Start writing about your own family. How were you raised? Were you breastfed? If not, that will be the first experience you were powerless over. Where did you sleep during the first year of your life? Where were you in the birth order in your family? How many siblings did you have? Was your family well off? Did you have your own bed, or did you have to share a bed or bedroom with a brother or sister? Did you have sufficient food? Did you have to compete for food? Did you have to compete for attention in your family? What was your feeling when you didn't receive enough attention growing up? Feel the urgency that you felt then and try and connect that to your present day need for attention. Try and connect these feelings to your survival needs. Did your oldest brother get all the attention? Notice how you feared not thriving like he appeared to be? Did your older sister receive more attention from mom and dad than you did? Did you feel envy? Anger? Did you fear you would never be as successful as she, that she'd always get the more desirable boyfriend? Again, try and connect these feelings to your fear of not surviving emotionally and relationally.

Now try and connect these personal feelings to the feelings our ancestors must have felt hundreds and thousands of years ago. At the beginning of our development as a society, competition was necessary for survival. In a group of young children in pre-historic times a child who was unattended to was at risk for being snatched by coyotes or dingoes or hyenas. Try and imagine what it was like for your ancestors. How did they live? Now try and connect your own feelings as a child with the feelings of your ancient ancestors. Become aware of the primal necessity of your feelings today and notice how at a deep level you are connected through your feelings to primitive times. Realize that your brain does not know you live in a world of plenty; your brain believes you live in a dangerous, world of scarcity. Lastly, notice how those ancient fears of survival manifest in your thoughts, feelings, and actions in your adult life today. Become aware how that ancient powerlessness became ingrained in your unconscious and structured your behavior, perceptions, thoughts, and emotions and those of almost every other human being.

Continue writing about how you were powerless over competitiveness and how it was passed from generation to generation through the socialization practices of our society. Write about your own aversion to cooperation. What are the limits of your ability to cooperate with others. If you were involved in sports, notice how you can cooperate with the others on your team, but you compete with everyone else. Notice further how competitiveness is enshrined in all of our sports and most every area of our lives. Become aware as well of the place that trust and mistrust occupy in our individual lives and in our society. Connect your own experience with competition to your impulse to control all situations, all outcomes.

Examine your impulse to accumulate more, more money, more property, more things, more possessions. Notice your experience of never having enough of anything, never being satisfied. Be willing to look at your emotion of greed, the emotion that motivates your acquisitiveness. Write about your powerlessness over greed. Where did this come from? Where did you learn to want more and more and more? Who taught you? Who modeled this for you? Underneath greed lies a primal dissatisfaction with what you have. Is it inadequacy that drives you? Do you believe you will finally "measure up" if you have more? Write about your experience of having difficulty accepting just what is. Ask yourself what would you do with your time if you were not driven by greed and the need for more. What is missing that you are trying to make up for by accumulating.

As you write about greed and the compulsion to accumulate, notice the challenges you have around accepting and believing that the universe is providing for your needs. For example, suppose you have a job and you can make ends meet on what you make. You have time for your family. You have time for a hobby. What motivates you to try and move up the corporate ladder or to become a supervisor or a foreman?

Several personality characteristics often accompany the drive to accumulate. Remind yourself that the first sentence of the book told you reading this book was not going to be easy. Now, examine your attitude of entitlement, your belief that you deserve to have more than others. In the pursuit of more, have you manipulated others? Write about it. Have you been impulsive? Have you been aggressive? Where did you learn aggression? Society conditions citizens to be aggressive, to forcefully take what they impulsively want.

How have you felt about the people who were hurt or harmed in your pursuit of more? What is the quality of your empathy?

One of the core qualities of the Contagion-infected person is narcissism. Write about any beliefs you have that you are better than others. Do you fantasize about power, success, or how attractive you are? Write about it. Write also about times you might have exaggerated your achievements or abilities. Do you expect praise? Do you believe you're special? Write about it. Do you recognize other people's feelings and emotions? If you have difficulty with awareness of the hurt others may feel or the shame, you might have narcissistic qualities. How do you treat people whom you believe are inferior to you? If you treat them with disdain, you might have narcissistic qualities. As you write your responses to these questions you may become aware that you have deficits of conscience. That would not be surprising. Realize that the imminent death of our world is a direct result of people believing they are superior to nature, that they are powerful and deserve to get their way. Our world is dying because of narcissism. People have lost the ability to feel empathy for the death of whales, for the destruction of the environment. As you write your responses, difficult as it may be, notice that you are developing insight into who you have been and the effects of who you have been on the world you live in.

The qualities we have been examining often express themselves as bigotry—racism—feeling superior to other cultures and ethnicities and religions. They can also express in hatred of women—misogyny. Use this opportunity to honestly write about your attitudes toward other races, religions, and ethnic groups. Examine your behaviors, your thoughts, and your emotions. For men, write as well about your attitudes towards women. Do you feel superior? Do you feel you have a right to be better paid, to have more authority? Do you feel you have a right to sex?

What are your attitudes towards reproducing? Have you thought about having children? Is it something you will plan for, or will you just "let it happen"? Thinking back over the chapters in this book, what comes up for you when you think about bringing children into a world that is dying? Is this fair to the children? Ask yourself what motivates you to have a child or children? As you write, raise your consciousness and become aware of your actions and motivations.

We humans have always been powerless over the natural world. Earthquakes and tornadoes and hurricanes have demonstrated how little power we have over the physical world we inhabit. We have lived for the last 25,000 years since the most recent ice age in a period of relatively benign climate. Now, due to our own actions, we are learning just how powerless we are over the natural world. We're finding out that we cannot burn all the fossil fuel we want, catch all the fish, or dump our plastics and waste into the ocean and the earth. Try and bring these issues to a personal level. If you live on a coast like Florida, write about how you are powerless over the rising oceans. If you've been hit by a hurricane, write about how powerless you are over the decisions made by corporations and politicians that have produced the climate change that has contributed to the more intense weather events. Write about your powerlessness over the melting of the Arctic and Antarctic ice and the release of methane. Write about how powerless you are over the increasing planetary temperature.

Write about your powerlessness over the political decisions that affect our lives and our survival. Admit your powerlessness to affect public policy. Acknowledge that the situation on our dying world is an emergency that requires immediate, forceful action; admit that you are powerless as an individual to do anything about it. The glaciers are melting faster than Congress will move. The world's governments have issued report after report and yet nothing substantive is being accomplished and we humans are individually powerless to change that. Write how you feel about the failure of governments to rein in the corporations of Big Energy.

Lastly, write about your powerlessness over death. Write about your fear of death. The world is dying, and we are powerless to stop it. Collective action is required of all the people on the planet, and no one is acting collectively. A globe-spanning death process has begun. Write about your powerlessness to influence or affect that process.

Unmanageability

Unmanageability comprises the consequences that arise from our powerlessness. Because we are powerless over greed and reproduction and competition and violence and all the others listed above, these outcomes have happened.

Thinking about your very early experience; if you were not breastfed what have been the effects? For example, do you have difficulty feeling safe being close to others? If you are a man, how has not being breastfed affected your relationships with women. Are you more likely to feel unmattered in relationships? If you are a woman, do you have difficulty being close to other women? Were you safe in your early relationships? If not, how has that affected your ability to trust others? If you were harmed by the people who were supposed to care for you, how has that affected your choice of partners? Develop an understanding that how you were treated in your early life becomes a model for your patterns of thinking, feeling, acting, and relating in your life after adolescence.

Now shift your awareness from your personal unmanageability to the unmanageability facing our world and us humans. Society's unmanageability is an accumulation of the individual unmanageability of all of the members of society. Many humans are motivated by acquisitiveness to some degree. For some the desire to have more—more money, more property, more power, more influence—has skidded off into greed made worse by manipulation, competitiveness, conflict, entitlement, aggression, selfishness and when these infect the society as a whole, the result is climate change, global warming, melting of glaciers and ice sheets, pollution, and ultimately the Sixth Extinction of the anthropocene that we are living through presently. Climate change and the attendant global warming are consequences of the Contagion. All the consequences enumerated in previous chapters comprise what Step One calls "unmanageability." As a result of Contagion all life on Earth has become unmanageable. Write about these consequences. How have your own fear and uncertainty and the fears and uncertainties of people just like yourself translated into growing defense budgets? Have you fallen for the paranoid rhetoric that insists we are under attack and must defend ourselves by controlling as much of the globe as possible? Write about how the bloated defense budget deprives society of resources to provide universal health care, education for all, 3-day work week, and income equality.

Wars, mass shootings, gunshot deaths, and murders are social manifestations of individual fears of loss of control that result in paranoid fantasies. People often feel powerless and their feeling of powerlessness gets compounded. They feel powerless over feeling powerless, and they succumb to advertising and social pressure and purchase a weapon. Their fear and insecurity causes them to dismiss or deny the facts that a weapon in the house will most probably be used to kill someone in the house or a child or a child who is visiting. When that fear infects the society as a whole, wars are likely. Weapons makers and politicians can easily manipulate the society by using their fears. Fear eradicates conscious thought and makes it difficult to imagine alternatives to buying a weapon for the house. Fear causes people to accept the NRA propaganda and oppose background checks for gun purchases.

Fear of loss of control includes fear of losing one's identity, which is the basis for misogyny and mistreatment of women. Women who were abused in childhood by the adults who were supposed to care for them have difficulty recognizing and avoiding men with fragile self-concepts who are most likely to abuse. Men who were socialized by parents who fought and fathers who abused have often internalized what they learned in childhood. Raised with fathers who used aggression and violence to get their way, they never learned emotion regulation, and so when they feel their identity is threatened they attack verbally and sometimes physically. Men with weak self-concepts often use criticism and verbal assaults to control their female partners and to keep them subservient so they never threaten the male's identity.

As you're examining identity, write about your unexamined and quite possibly unconscious compulsion to have children. Obviously your parents had a child or children. Examine how they modeled having children as a symbol of being an adult. Now think about and write about how your compulsion to reproduce became the overpopulation that our world is facing now. As human population has ballooned, we have moved into habitat that formerly was home to many species of animals that are becoming extinct. A primary cause of species extinction is habitat loss. Species extinction is a consequence of our powerlessness over our impulse to reproduce; species extinction is part of the unmanageability arising from the Contagion.

Overpopulation leads to deforestation and depletion of the oceans' fisheries because more and more people require more and more food, and more and more building materials and more and more land for growing food to feed the cows that humans turn into food. Because there are more and more people, more and more factory fishing boats harvest more and more of the oceans' fish and the fisheries can no longer sustain themselves. Overpopulation leads to depletion of resources like forests and fish. Write about how the urge to reproduce, which you feel as a personal imperative, has collectively become overpopulation. Overpopulation is one element of unmanageability.

We humans have externalized our competition, acquisitiveness, and greed in the corporations and bureaucracies we've created. The corporations and bureaucracies have developed an independent life. They have power and control over human beings. The corporations and bureaucracies seize on resources like oil and natural gas and land and minerals, develop them and make products that they use to create profits for themselves by selling manufactured goods to the people. The corporations together with their co-conspirators the bureaucracies have destroyed the world's environment through oil and mineral extraction. Fracking and mining have ruined water supplies, contaminated soil, and polluted the air we breathe. In order to increase yields, agricultural corporations and their associates in the chemical industry have synthesized and sprayed herbicides and pesticides on the food we eat.

The bureaucracies of government aid and abet the corporations by weakening oversight and regulations. We humans are powerless over the motivation of profit that drives business. The pollution and contamination and environmental destruction comprise the unmanageability that derives from our powerlessness over the Contagion-driven pursuit of profits. Write about your personal experience of the environmental degradation. Do you live in Flint, Michigan? Has your drinking water been poisoned with lead? Do you live near a fracking well? Is the water coming out of your tap flammable? What poisonous or carcinogenic chemicals are you forced to drink because you are powerless over the corporations and bureaucracies that are compulsively making profits?

Life on earth has become unmanageable. Climate change and global warming are worsening. We humans and much of life on Earth are faced with extinction. The Sixth

Mass Extinction of the anthropocene geological age is the unmanageability that has arisen because we are powerless over the Contagion.

As you have read these sections on powerlessness and unmanageability have you noticed yourself resisting the statements? Have you thought it's not as bad as that? Have you thought that we humans will find a solution because we always have? This is your denial. Denial is a powerful component of our unmanageability. Extinction means death and many people fear their death. They are stuck in the denial phase of grief. Write about your denial. Write about your resistance. Become aware that most of the members of our species are similarly stuck in denial. Write your thoughts and feelings.

Reservations

Reservations are a component of unmanageability. Reservations are the cognitive defenses our brains create to prevent us from acknowledging our powerlessness and unmanageability. Reservations also tell us that we'll try the Humanity Rising Program but if the environment doesn't improve or the oceans continue to rise then we'll go back to our old, Contagion-driven ways of behaving. Reservations prevent us from completely surrendering to the program. Alcoholics, for example, may tell themselves that they won't drink *unless* their spouse leaves them, and if that happens they'll have permission to drink again.

We may tell ourselves that we'll admit to powerlessness and unmanageability until or unless some event happens. Write about what that event would be for you. Loss of your job? Illness? Divorce? Whatever it is, if or when it occurs we'll have permission to return to competing, to acting greedily, to overusing resources, to violating the commons, to controlling our wives, to murdering our neighbor, to attacking another country. We tell ourselves that if we cease our greedy behaviors, then the corporate executives and politicians who do not cease their greedy behaviors will take over even more of the world.

Step Two

Came to believe in a power greater than ourselves that could restore us to sanity.

Step Two is critically important to recovery from what Step One showed us that we are powerless over. Self-centeredness, selfishness, and greed are fundamental components of the Contagion. The person dominated by Contagion thinks only of himself or herself. Self-centeredness is the core of the cancer of Contagion. It is a character defect that leads to all the other character defects. The Contagion prevents a person from serving All Our Relations, prevents a person from serving the world. The Contagion precludes one from serving any power greater than the self. The Contagion prevents one from working with others, prevents cooperation and mutuality. Contagion deceives a person into serving a false higher power like a corporation or bureaucracy or a political party.

Insanity is a key concept in Step Two. It is insane to separate oneself from the family, clan, or tribe and to take actions that harm other human beings. Racism is one of the insane behaviors that arises when people separate themselves from oneness with other human beings. Bigotry is an insane behavior. It is insane to separate ourselves from oneness with the world. When people lose a sense of oneness with water, air, and earth, they believe they have power over the natural world and then they can act out their self-centeredness on the environment. Pollution of the air and water that sustain us is insane. Contamination of our food with herbicides or pesticides is insane. Pouring CO_2 into the atmosphere and warming the planet to where the glaciers melt and oceans rise is insane. Believing we humans are separate from nature, from All Our Relations is insane. Insanity is a central element of Step Two. The Contagion IS the insanity. Repeating greedy, self-centered, fear-driven, Contagion-driven behaviors over and over again and expecting a different result is the definition of insanity.

In spite of all the teachings of religion, the power greater than the self is not "god." "God" is a projection of the self. "God" takes our fear of death and turns it into a false promise of everlasting life. The churches interpret "god" to say the insane acts we commit are "god's" way of testing us, giving us suffering in order to lead us to "god." Contemporary human suffering derives from the actions of human beings, not the actions of "god." The god-concept allows humans to avoid feeling shame and guilt for the damage we have done and are doing to this beautiful world and the other humans, animals, and plants we share the world with.

Christianity promised eternal life without any evidence. Christianity's promise of eternal life dissociated Europeans from an integrated life with plants, animals, other humans, the physical universe and all of the world. Europeans did not believe they would die. They fantasized that although the body would die the "soul" was immortal and would ascend to heaven upon the body's death. The European's developed a clever religious workaround to avoid the terror of death. As we have seen, Europeans have externalized the fear of death through murdering others. Apparently the idea of an afterlife has not resolved the terror of death.

Religion transmits a central theme of Contagion, the idea that some people are special and that their specialness is conferred on them by "god" and that the world is divided into special human beings and not-special human beings. Specialness is insane. Religion served the Contagion. Religion preached that some human beings were heathens, unclean people, because they had not been baptized. Religion supported and allowed the conquerors to steal the lands and property and treasure and lives of those human beings designated as pagans or infidels. Any religion that divides human beings into groups where some are "saved" and some are heathens, any religion that allows or promotes murder, slavery, or war serves the Contagion and is insane.

Write about the religion you were raised with. Write about how religious beliefs and attitudes were installed in your unconscious at an early age. Write about how difficult it is for you to consider changing those unconscious schemas. Write about how important the ritual of your religion is. Ritual can be separated from the beliefs of a religion. Rituals embody a "participation mystique." The Muslim ritual of praying five times a day while facing Mecca joins the male—women are either excluded from the prayer hall or segregated into a separate space—with other Muslim men. The Christian ritual of communion joins the participant in a group process. Write about the exclusionary practices of your religion. Consider your own attitudes about LGBTQ people and how those are the same or different than the teachings you received as a child.

Nature, the universe, can be a power greater than the self. When Nature or the Universe is a power greater than the self, we humans are then an integral part of the power greater than the self. Not separate from it. Not given dominion over it. Not allowed to diminish it or use it solely for our own purposes. When we are at one with Nature and Nature is our higher power, we realize that the water in our bodies is the same as the water in rivers, lakes, and the ocean. When we realize this, we immediately realize the insanity of contaminating the oceans, lakes, and rivers with harmful chemicals. When we are one with Nature, we realize the air we breathe is the same air breathed by all other humans and every living thing. We realize the insanity of the kind of ownership that allows one company to contaminate the air we all breathe. The Contagion of self-absorption allows humans to contaminate the air, water, and land that we all rely on for life. Believing that Nature is a power greater than the self begins to heal the insanity of Contagion.

Insanity and the Contagion are driven by death anxiety. Fear of death motivates greed, because the greedy person believes he can never have enough resources to keep him alive no matter what. Step Two asks you to seek a belief system that integrates life and death without inventing imaginary places like heaven, purgatory, and hell that have the function of assuring eternal life. As unpleasant as hell is supposed to be, the sinner is still alive even though suffering eternal torment. It is noteworthy that Contagion, manifesting as religion, accompanies death anxiety wherever death anxiety is found.

Belief systems do exist in which death is accepted as an integral component of life. Indigenous people's world-view and the rituals derived from it integrated death. Indigenous spirituality provided for death. Tibetan Buddhism was preceded by an Indigenous, shamanic belief system called Bön that like Native spirituality integrated life and death. Tibetan Buddhism integrated Bön teachings, and Tibetan Buddhist masters understood that a life of spiritual practice is a preparation for death. The Indigenous spirituality was an earth-based world-view, and today's Gaia concept is not materially different from the Indigenous reverence for Our Mother the Earth.

The currently unfolding world-wide extinction disaster derives from the fact that Europeans and their descendants have no world view that integrates death into the life process. Step Two is based on the belief that the act of believing in a power greater than the self will promote recovery. Write about your thoughts and feelings as you strive to find a belief system that frees you from death anxiety. Are you willing to explore and investigate different belief systems than the ones you were indoctrinated into? What are your thoughts and feelings about death? How have you attempted to suppress or otherwise deal with your thoughts and feelings about death?

Step Three

Made a decision to turn our will and our life over to recovery from Contagion

The phrase "our will and our life" refers to our thoughts and our actions. The words "conscious decision" refer to the theme of awareness that runs through this book. Through these steps we are raising our Contagion-driven thoughts and actions to awareness where we can consciously choose to change them. Step Three asks us to ponder what those thoughts and actions are that have driven us to be participants in the destruction of our world. Are we willing to change them? Humanity Rising is a protocol for changing cognitions, emotions, and behavior. It is also a spiritual program. As you think about your Contagion-driven life, ask yourself if you are sick and tired of being driven by greed, avarice, mendacity, anger, fear, entitlement. Are you sick and tired of being sick and tired? Humanity Rising offers you a respite, a new way of living. Write about your openness and willingness to turn your will and life over to a new way of living.

In Step Three we make a conscious *decision* to turn our will and our life—our thoughts and our actions—over to recovery from Contagion. This is a *decision* step. In succeeding steps we will take *action* that manifests our Step Three decision. In writing about Step Three, examine what it means to you to make a conscious decision to turn your thoughts and actions over to Mystery or the Universe. Examine how you might think differently or emote differently. Allow yourself to wonder how differently you might act. What different attitudes might you have?

Step Four

Made a searching and fearless moral inventory.
This step asks us to consider who we were at the moment of our birth. It is at this point and maybe only at this point in one's life that we are most free of Contagion. When we are free of

Contagion we are closest to embodying our moral inventory. At the moment of birth we have not learned to lie, we have not learned greed, we have not been exposed to violence and learned to be violent. Nobody has yet taught us entitlement or misogyny.

Every child lives in an environment where the necessities imposed by reality interact with an ideal of how the world should be. The infant born into an aboriginal world may actually have had a life closer to the ideal in some respects. From birth until the child was capable of walking, the young person was carried in a sling by mother. Mother was never far away or detached emotionally and physically from the child. Psychologists have discovered over the last 50 years just how important closeness to mother is to the child's healthy development. The modern world begins to teach Contagion almost from birth. If mother has to work and leave the baby, the attachment bond between mother and infant is disrupted. Breaking the attachment bond conditions the infant's forming brain to a state of lack, of want, of insufficiency that will affect how that person acts throughout life.

The concept of a moral inventory may be challenging. For people unfamiliar with the idea of a moral inventory, it may be easier to find your moral inventory by examining your immoral inventory. Write about your carbon footprint. How much gasoline do you consume? How many cars do you own? What have been your attitudes to toward your consumption? If you realize that you did not care about your carbon footprint, that is your immoral inventory. Your moral inventory will be the opposite: caring about the environment and caring about your use of scarce resources. Do you feel entitled to a new car every couple of years? That will be your immoral inventory, and your moral inventory will be acceptance of the sufficiency of what you have. What about money? Do you spend your time plotting how to make more money? Notice how the Contagion drives you. Write about it. Your moral inventory will be acceptance of sufficiency. Your moral inventory will be the mirror image of your immoral inventory.

Make a list of your moral inventory items. Examine every area of your life. How does your relationship to women or to your wife manifest your immoral inventory? Do you feel entitled to order your wife or your children around? Entitlement will be your immoral inventory. Write about your motivation to have your wife and children act the way you want them to act, to accomplish what you believe they should accomplish. Now ask yourself what is the opposite. Your moral inventory will be acceptance that your children are individuals with their own thoughts and feelings. Your children are going to model their adolescent and adult behaviors after what they learn from watching and copying you. If you want them to be kind, you have to model kindness. If you want them to respect each other and other children and adults, you must be respectful.

What have been your attitudes and behaviors toward the natural world? Have you felt entitled to kill animals for sport? Have you felt entitled to tear up the desert or the wilderness on your ATV or dirt bike for your enjoyment? Your moral inventory will be loving compassion for the desert and wilderness. What about trash and garbage? Have you felt entitled to just throw beer cans and plastic bottles and fast food packaging away wherever you felt like? That's your immoral inventory. Your moral inventory will be respect. It will be an attitude of gratefulness and oneness and mutuality with this beautiful world we live in. As we work these steps we learn to prayerfully intend that we will live in our moral inventory.

You do not have to associate with Contagion-infected people, just as an alcoholic does not have to drink in a bar just because other humans are drinking in bars. However deeply ingrained your Contagion behavior, you can change. As you work the Steps and raise your behavior to awareness you will have the choice to act on your moral inventory or not.

Step Five

We admitted to our higher power, to ourselves and to another human being the exact nature of our wrongs.

In Step Five we honestly admit to ourselves, to another human being, and to our higher power all the ways that we have violated our moral inventory. Review Step Four and pull out all the thoughts and actions that were selfish, critical, angry, entitled, greedy, self-centered. Write these down. Let's suppose you were critical of your wife, the mother of your children. You don't have to write down every instance of your belittling her or interrupting her or talking over her or controlling her. Writing three is enough. Write enough so that you begin to feel the pangs of shame at your behavior. Did you criticize and shame your children? Were their grades never enough to satisfy you? Those are wrongs that you need to admit. Write them out.

Thinking about the environment, write about how you used scarce resources because you could. How did you contribute to pollution? Did you invest in energy companies knowing that they fracked? Did you buy stocks in companies that spilled oil? Did you fly on commercial flights because you could without thinking about your carbon footprint? What excuses did you make for your behavior? Your justifications and self-serving explanations will be part of your Fifth Step. Your entitlement will be part of your Fifth Step. If you felt resistance to answering these questions, write that out because your resistance is your immoral inventory and demonstrates your violation of your moral inventory item of acceptance.

Did you kill animals for sport? For trophies? Did you drive off-road in fragile desert or wilderness? Did you swerve your car or truck to purposely kill squirrels who ran in front of you? Write these down and then think deeply and write about the attitudes that motivated you. When you have written out your Step Five, meet with your sponsor and read your Fifth Step to him or her.

Steps Six and Seven

We became entirely ready to remove these defects of character

Humbly set out to remove our shortcomings

Following Step Five you will have an understanding of your character defects. Steps Six and Seven take us deeper into the recovery process. Steps Six and Seven require willingness and surrender. We begin the process of shedding the old ways of being that have not served us or our world.

Through an honest Fourth and Fifth Step we ourselves identified our character defects: our greed, our acquisitiveness, our competitiveness, our desire to control, our misogyny, our fear of death, our demand for power and control. Having identified them, we must change them. This is the work of the program of Humanity Rising. Raise the character defects to awareness, notice them when they resurface, and then choose to have different thoughts, different emotions, and different behaviors.

We changed our personalities. AA's version of Step Seven says we humbly asked God to remove our shortcomings. Humanity Rising has eliminated the reliance on a god. There is no "god" to remove our character defects. As the book has argued, the concept of god has been a major contributor to the Contagion. As we shift our world-view through Step Two to an understanding of our oneness with every living thing, we begin to understand that the world itself and all our relations, the sun and solar system, the galaxy and universe comprise our higher power. Because we are one with our higher power, we have the responsibility for removing our character defects ourselves. You may seek help. A psychotherapist may be able to help, or sponsor. Ultimately change is your responsibility. If you have been controlling with your wife and children, you will need to become aware of your controlling behavior and change it. Choose to do the exact opposite of what the Contagion was wanting you to do.

The most effective way to remove an undesired behavior is to replace it with a desired behavior. In Step Five we identified our Contagion-driven behaviors, thoughts, attitudes, and feelings. Working with a sponsor and through discussions with others we begin to change what we can. If we are greedy, we notice it and substitute caring and giving thoughts and behaviors. We replace materialism with satisfaction and gratitude for what we have. If we are motivated by avarice to accumulate more money and more property, we notice those thoughts and emotions and we replace them with kindness, altruism, and giving. We try to be aware of the drivenness to spend long hours at work striving to accumulate more and more money, property, and stocks. We notice what the cost has been to our wives and families. We make a decision to take different action. Perhaps we attend our daughter's ballet practice or our son's football workout. Notice how different you feel when you are mattering your children, interacting with them, listening to them.

Humanity Rising hosts on-line and in-person meetings. Meetings provide a place to talk about the challenges of living a life free of Contagion. In a meeting a member is assured of confidentiality and anonymity. We introduce ourselves by our first name only. If we have something to talk about but don't want to share at the meeting level, we meet with our sponsor one-on-one. In order to sustain the group, we participate by taking service commitments: literature, chips, refreshments, secretary, and representing the local group at the area level. We model recovery by collecting time free of character defects. We sponsor members. Local meetings can have themes that organize the group's focus. One type of meeting is the death awareness group. In a death awareness group we discuss death and participate in a rational discussion of death. We do not offer advice or comments to another member. It is very helpful to be heard, to be mattered. Death is a forbidden topic in society, and as a result we suffer from having to bottle up and contain our thoughts and fears. As we have seen, when death is not acknowledged it is externalized.

Steps Eight and Nine

Made a list of all persons and things we had harmed and became entirely ready to make amends to them all.

Made direct amends to such persons and things wherever possible, except when to do so would injure them or others.

To make an amends is to compensate for an injury to another, to atone for a wrong. In Step Eight and Step Nine we make amends to the people, cultures, and environment we have harmed. A Ninth Step amends consists of an outer amends and an inner amends. The outer amends is the overt action of contrition. Admitting the behavior and apologizing for it. If money has been stolen, we return it. If property has been taken, we restore it to its rightful owners. The inner amends comprises the changes we make within ourselves, changes to our thoughts and feelings, that are sufficient so that we do not repeat the behavior.

As the book has argued, people of European descent owe an amends to the Indigenous People whom we murdered and whose culture we destroyed. Columbus brought Contagion to the western hemisphere in 1492. The early settlers brought Contagion to what is now America in 1620. Driven by Contagion the settlers killed and slaughtered with impunity. Although it occurred 400 years ago, the murderous actions of the early settlers have left a stain on our national soul. This stain has never been addressed; no amends has ever been made. We live with the consequences of our failure to atone for the actions of our ancestors. America's militarism, America's fear that someone will invade us, is a direct result of our failure to atone for our actions. Because we have not made amends, we perpetuate the Contagion in all the ways that have been described.

Requiem proposes a national year of atonement as an amends for the genocide Europeans perpetrated on the Indigenous people. Canada carried out a successful program called the Truth and Reconciliation Commission that addressed the effects of sending Indigenous children to the reservation schools. To free ourselves from the consequences of Contagion, America must engage in a nation-wide, all encompassing national year of truth and reconciliation to Indigenous People. America's schoolbooks do not tell the truth of the genocide that European settlers enacted on the Indigenous people. The textbooks must be changed. American leaders must at last actually lead and stand before the nation and admit honestly that we stole the homelands of the Indigenous people and murdered them. Americans must participate in meetings at the local level where we engage in honest dialogue with Native people and with each other about what we did. We invite Indigenous People and listen to their anger and anguish without cross-talk. They have never been able to tell their true story and to have it heard. The truth must be told and it must be heard. We ask to hear the thoughts and feelings of the oppressed. The descendants of the first people who were living here when Europeans first arrived must tell what their lives are like now and how that relates to the lives and land that were taken from them. Americans of European descent must listen without judgment, without arguing or contradicting. Without defensiveness or justifying. When we have heard the descendants of the first people, we must make amends to the Indigenous people whose culture and lives the invaders destroyed. As we process what was done, we descendants of the invaders must admit to the benefit we have received 400 years later from the massacres our ancestors perpetrated.

We examine how we have continued to oppress people of Indigenous descent even today through denying health benefits, through denying birth control, through denying abortion, through steadily eroding what little land we did cede to Indigenous people by for example trying currently to seize the lands of the Wampanoag for a casino in Cape Cod. We admit that through our Big Energy corporations we have continued to take land for drilling and to contaminate the land and water of Indigenous people for our benefit. We have not listened to the protests of Indigenous people. We have arrested them and sicced our dogs on them and beaten them. We have refused to be moved by their songs and their complaints.

As we descendants of the invaders begin to realize the enormity of what we did we begin the process of atonement. Descendants of the first people are still trying to live on reservations where they can no longer practice the old ways, the ways they were living when Europeans arrived. We must find ways, create programs to return some of their lands to Indigenous people. We must stand before our own people and our Indigenous neighbors and speak our amends. Perhaps we will give gifts; or create art work. Perhaps as we learn to communicate honestly with each other we will create a ceremony or ritual that embodies the healing and reconciliation we are sharing.

Steps Eight and Nine are a process, not a single event. Having started our amends with the first victims, the Indigenous people, we continue our amends with a national year of atonement, a national year of reconciliation to African American people. Slavery in America antedates the arrival of colonists at Plymouth. The first black slaves arrived in Virginia in 1619, brought by Dutch traders who had seized the 19 slaves from Spanish slavers. Dutch and Spanish slavers introduced Contagion to Africans. Soon tribes infected by Contagion were capturing and selling people from other tribes to the slavers. Africans also sold people from their own tribes. In earlier chapters, *Requiem* has enumerated the violence perpetrated on slaves. The violence and the violation of personal freedom have stained America's soul. The violence continues. Because Americans have never made amends for what their ancestors did, the stain continues to affect our nation 400 years later. Ignoring the violence and violation has not removed the stain. The stain affects the nation's relations to blacks, it affects relations of blacks to blacks, and because it stains the unconscious of the descendants of the slavers and slave owners, the stain continues to affect relations of whites to whites and whites to blacks even today. Because the descendants of slave owners have not made amends to blacks, the stain of slavery expresses in hidden ways.

Slavery creates an underclass, and the principle of an underclass has been institutionalized in American life. Women are treated as an underclass, politically, socially, and in relationships. The white male descendants of slave owners prevent women from making the same wage as men, which perpetuates women as an "other," unworthy. The white male descendants of slave owners also expect women to work like slaves, insisting they hold two jobs and also fix meals, shop, clean, and raise the children. Sometimes the entitled white male privilege spills over into spousal abuse. A process of atonement and reconciliation can begin to heal the social and personal consequences of slavery.

We express our amends through meetings and gatherings we organized where we admitted how we today have benefitted from slavery. We admitted how we have benefitted from creating an underclass. We admitted how we have benefitted from creating an "other" a class of people that we could project our own "badness" on to. We admitted how we have continued to enslave black people through denial of medical care, through shootings, through gerrymandering, through restricting the freedom to vote. We listened to what black people had to say. We admitted to our fear of black people's rage. We admitted how we have benefitted from white privilege.

The stains of genocide and slavery corrupted the souls of Americans and manifested three centuries later through abuse of the environment. The water, earth, and air have become "other" just as Indigenous people and African Americans were treated as "other" and abused. The Contagion affected how we treat the environment. Human beings affected by the Contagion have not been able to subdue the environment as they had subdued Indigenous and African American people. The unmanageability of our powerlessness over the Contagion is manifesting in climate change, global warming, and pollution of water, air, and land with disastrous effects. The Sixth Extinction is a direct result of the Contagion and its effects on human thinking, feeling, and acting. The Indigenous people tried to fight back against the invaders, but they were out-gunned. Blacks tried revolts. Nat Turner led a revolt in Virginia during which about 60 human beings, mostly whites, were killed. The revolt was brutally suppressed. In the retribution that followed, many innocent blacks were killed as were local Indigenous people. Turner was eventually captured, tried, sentenced to death, hung, and his corpse drawn and quartered. The climate and the earth are fighting back effectively against the Contagion. What Turner could not accomplish, the Sixth Extinction will.

As members of a nation and a system that has brought about such extensive climate change, pollution, and degradation, we owe individual and collective amends to the environment for the destruction we have caused. *Requiem* proposes a national year of atonement and reconciliation to the environment, to the plants and animals that have died or gone extinct because we shot them, poisoned them, trapped them, cut them down, plowed them over. Steps Four and Five have shown you your own experience of greed and disrespect and hopefully connected you with all the others who as a group have supported destruction of the environment for personal gain.

The governmental bureaucracy at all levels has acted on behalf of the collective and expressed the national greed through seizing lands given to the Indigenous people and through giving Big Energy and Big Mineral corporations the rights to extract oil and ore from national parks. The goal of recovery from abuse of our mother the Earth is a complete change of behavior, thought, and feeling. As part of making amends to the environment, America can expand its parks and wilderness. The nation can firmly and without exception remove the corporations of Big Energy and Big Mining from the national parks and monuments and sacred lands where they have insinuated themselves. Making good and appropriate use of the U.S. Forest Service, the nation can plant many varieties of trees to replace the trees cut down, to restore the forests the have been mutilated. We express our amends to the forest environment by bringing back the diversity that had been eradicated when we converted the native forest to tree farms for Big Lumber.

Making amends to the environment entails undoing all the laws and regulations that have given corporations power to operate America and much of the world for the corporations' benefit. The world economy *uses* people, it does not benefit them. All the monetary benefits flow upward to the 1%. The American and the world economy function like a coal mine where the profits all go to the wealthy owners and where the workers labor 10 hour days and six day work weeks and get paid with Black Lung Disease for their work. In the American economy—Europeans have made changes in health care and education that benefit them in ways America has not—workers labor for low pay, work without health benefits, are denied free higher education, do not make enough to own a house, and see their retirement benefits stolen by a malignant Congress. The nation can begin to make amends by taxing the obscene wealth of the One Percent. The nation can rein in the military-political-industrial complex and reduce the defense budget to what it was in

1960: $344 billion. The nation can invest its profits through taxation to provide universal medical care, free higher education, and an inviolate retirement system. When the needs of the people are taken care of, the people will have motivation to heal the nation.

An effective Ninth Step makes amends to the environment and addresses individual wrongdoings as well as social and political shortcomings. *Requiem* has discussed consumerism. Your Step Five should have helped you investigate your consumer behaviors. How have you made amends for consumerism? How many cars do you own? How many properties? What amends through positive action can you make for consumerism in your life? Do you receive income from investments? Do you have rental properties? Write about the rents you charge your tenants. What is the effect of the rent you charge on your tenants' ability to feed themselves and their children? Do you care? Write about your feelings about your tenants and their difficulties. If you come to realize you are charging your tenants a usurious rental, what would it be like for you to reduce the rent as an amends to the renters? What feelings would come up for you? What changes of personality would you have to make to work with your tenants for the mutual good rather than your own selfish needs? Write about this. What organizations could you support that would manifest your change of behavior, thoughts, and feelings toward the environment?

Step Ten

Continued to take a moral inventory and when we were wrong promptly admitted it.

Step Ten asks you to examine your behavior on a daily basis and assess whether you have lived in your Step Four moral inventory. Step Ten is a maintenance step. Step Ten can be worked out of order. Many members of Humanity Rising conduct a Tenth Step inventory at the close of every day. If your assessment shows you that you have violated your moral inventory, that you have slipped into your immoral inventory of greed, entitlement, and anger, you can immediately rectify the situation. As you work the Steps of Humanity Rising, you will have a better and better understanding of who you are and what motivates you. Step Ten allows you to work the change process in real time. You can begin to change immediately. You do not have to wait until you've completed all the previous nine steps.

Step Ten provides for course correction. Perhaps as you review your day, you will remember that you raised your voice and acted with anger towards a wife or child. As soon as you become aware of your Contagion-driven thoughts and behavior—perhaps you remember noticing the pain in the face and eyes of your wife or child, perhaps you recognize the shoulders slumping with shame or the body stiffening with defensive anger—you must stop, center yourself, and honestly admit to the wife or child that you have erred. You have relapsed however briefly. Over time as you incorporate these principles into your daily life, you will notice you have fewer and fewer slips. None of us are perfect. We will make mistakes. As we grow in the program will catch the mistakes more quickly and eventually we'll feel the impulse to act against our moral inventory *before* the words are said or the action performed. When that happens we will realize we are having the change of personality Step Twelve promises.

A Tenth Step inventory helps us notice when we're motivated by greed or when we've fallen into consumerism. Step Ten also helps us track our relations to the environment. We ask if we have recycled in all the ways possible. We can ask ourselves if we could garden instead of relying on store-bought food. Do you have a stock portfolio? A Tenth Step inventory can guide us in examining our investment choices. Often investors choose a particular company because of the promise of immediate monetary benefits. Using the Tenth Step process, you can research the extent to which the company externalizes the environmental costs of its production practices. Perhaps you were considering investing in a company that intends to drill for oil in the Arctic National Wildlife Refuge. Write about the choices you face. Discuss it with your sponsor or in a meeting. Ask yourself if you are maximizing profits—a Contagion-driven immoral inventory activity—at the expense of damage to the environment.

Step Eleven

Sought through prayer and meditation and ritual to improve our conscious contact with a power greater than ourselves seeking only for knowledge of how to better work the program of Humanity Rising and for the power to carry that out.

Tracing the concept of power through the twelve steps helps us clarify what it means to live in mutuality with the world and all our relations. In Step One we admitted our powerlessness over the Contagion that has driven our own lives and that is resulting in the Sixth Extinction. Step Two asked us to come to believe a power greater than ourselves could restore us to sanity. The word "power" does not appear again until Step Eleven when we prayerfully intend that our higher power will help us work our recovery program. Hopefully you will have recognized that the world, the universe, nature, or Mystery is your higher power. Hopefully, as you develop a connection with the world, a oneness with All Our Relations, you will realize that the trees and bushes of the forest are supporting you when they are valued and protected. Hopefully, as you spend some weekends cleaning the beaches of plastic and trash, you will realize that a pristine ocean beach gives you hope and power and you feel encouraged to continue with your efforts.

Step Eleven encourages prayer, meditation, and ritual. Humanity Rising suggests prayerful intention as the most effective form of prayer. When we engage in prayerful intention we are not asking a "god" to act on our behalf. We are stating our own intention for how we want to act. The outcome depends upon us. We acknowledge that we must exert effort to produce the change we want. When we meditate we open ourselves to receiving information from the world. In meditation one clears one's mind and allows receiving to occur.

Step Eleven also mentions ritual. A ritual is a ceremonial act, an observance. Humanity Rising distinguishes between religious and spiritual activities. Spirituality comes from the Latin word *spiritus* which means to lighten, to lift up, to fill up. Religion comes from the Latin word *religio* which means to bind. Religion is based on a contract between a higher power and the self. In Catholicism, people make a contract with God in which they are promised eternal salvation in exchange for obeying the Ten Commandments. Humanity Rising asks you to notice how you feel when you clean up an ocean beach. When you look back at the mound of trash you've picked up and bagged and carried to your truck, when you look at the pristine beach free of cans, paper, and plastic, do you feel lighter? Do you feel lifted up? Do you feel more at one with the beach and the ocean or lake? If so, you are having a spiritual experience. Treasure it.

In America and the West, we have many rituals but few that celebrate our relationship to a higher power. Rituals are often practiced in a group. Many of our rituals are profane meaning they are not concerned with something spiritual. They are secular. Secular rituals include watching sports on TV or attending a game. Raves are secular rituals. Drinking in bars is a secular ritual. Among the religious, attending mass or synagogue or mosque are forms of ritual. People who sing in choirs or other groups often experience the lightening that is characteristic of spiritual experience. Other forms of spiritual ritual include nature hikes, Sufi dancing, group meditation in a Sangha, sweat lodge, singing and dancing at a pow wow. Ask yourself how will you celebrate your higher power? How will you express your gratitude for receiving a new way of life, a new happiness? Step Eleven asks you to explore your beliefs about where your power comes from. Perhaps you once believed that having a six figure salary gave you power, or a large IRA, or a thick bankroll in your pocket, or a large stash. Step Eleven asks you to find or invent rituals that celebrate your relationship to your higher power.

Write about what your higher power was formerly. The problem with making a thing—money, property, IRA—your higher power is its transitoriness. An authentic higher power is lasting no matter what your circumstances may be. As you work the steps are you beginning to realize that you are one with the world and that is where your power comes from.

Step Twelve

Having had a spiritual awakening as the result of these steps we tried to carry this message to other human beings and to practice these principles in all our affairs.

A spiritual awakening is a change of personality sufficient to forego the addictive behaviors. The concept of "spiritual awakening" unites the domains of psychotherapy and recovery. Through the 12-Steps of Humanity Rising, we attempt to free ourselves of the addictive behaviors of the Contagion, and we also attempt to change our personalities, which entails a neurological brain change such that neural pathways that lead to desirable outcomes replace pathways that lead to maladaptive outcomes. This means changing our thoughts so that our actions change. For example, a substance-abusing addict may suddenly have a thought of using. In recovery, the addict immediately recognizes the maladaptive thought of using and replaces it with a recovery thought like, "There's a meeting in a half hour, I'll get over there right now." For an addict recovering from addiction to abuse of the environment, the thought "Hey, I recycled a cardboard box this morning I don't have to go to the trouble of finding a recycle bin for this soda can" is a Contagion thought. In recovery, the addict replaces the unwanted thought (which will lead to an undesirable action if followed) with the recovery thought, "C'mon, man. Recovery is thorough. If I don't recycle this can I open the door to abusing the environment." An addict recovering from greed might find him- or herself planning a major investment that will make a lot of money. In recovery the investor takes a deep breath and thinks about the consequences for the environment of the prospective investment. The investor asks if the investment will lead to more pollution or more global warming. The investor also notices his emotional motivation. He or she may consider alternative uses for the money, alternatives that would improve the world's environment. The investor may also consider it the prospective investment will enhance or diminish his or her spiritual condition. Quoting AA, "All we have is a daily reprieve which is entirely contingent upon our spiritual condition."

Every human being's spiritual condition is being tested now and will be tested more severely in the next decades. Based on geological science it is becoming clearer and clearer that there is to be no miracle. The world's death process is accelerating. Contagion has too deep a hold on too many humans. The corporations and bureaucracies have too much power. The early signs of extinction are evident in planet-wide warming, melting glaciers and sea ice, drought, vicious storms, methane release, and species extinction. These events will increase in frequency and intensity. Society will begin to break down. The social and environmental breakdown will try every human's spiritual condition. Contagion prevents the development of an adaptive spiritual condition. People infected with the Contagion will resort to the only skills they have: Contagion skills. They will become more greedy and more violent and more entitled and more possessive. Without a change of personality, they will die as they have lived: feeling want and desperation.

People who have worked the Humanity Rising program will enter into the acceptance stage of the grief process. They will accept that they cannot change the world's death process and they will grieve the loss of the world. They will celebrate the beautiful gifts the world gave them. Saying good-bye to the world is much like visiting a dying relative or friend in hospice for one last visit. Both you and the friend know you will not see each other again. You open your arms and hold each other. You feel the warmth in your heart that signifies your attachment for each other. If you were fortunate enough to have had a secure attachment to your mother, you will recognize this feeling. You can feel this same attachment for the environment, for the glaciers, for the forests, for all our relations. Remember, the environment is dying too. Support each other. Emotionally hold the environment and allow yourself to be held in return. What follows are some specific actions you can take that are appropriate to this Step. These actions will help you practice the Humanity Rising principles in all your affairs.

If you live on one of the coasts, go to the ocean one last time and sit in awe at its beauty and majesty. Allow memories of times past when you and perhaps your family visited the ocean. Allow sadness to arise for the plastic contaminating it. Make amends. "I apologize for the trash I threw into you." Collect plastic and trash from the beaches. Explore remanufacturing so that we move towards a society with zero trash, zero pollution, zero waste, a society where every single item that is manufactured is returned to its raw materials once it is no longer in use. Notice if you think that because the world is dying it is pointless for you to pick up trash or to make amends. Notice how this is a Contagion thought. Remind yourself that all you can do is live in recovery yourself. Others, seeing your actions, may be moved to change. Or not.

Revisit a favorite mountain. Hike a trail you've taken before. Allow your eyes to take in one last time the vistas that are disappearing. Listen to the sounds of the woods, the fields, the streams, the ocean. Express your gratitude to the mountain for the gifts it has given you. The smell of clean air, the aroma of trees, the wild animals who live there, the trout swimming in the lakes.

Sit beside a beloved river, one of the great ones— Columbia, Mississippi, Missouri, Sacramento, San Joaquin, Platte, Rio Grande— or one of the lesser ones. Recall times you spent there. Recall fishing trips. Speak your gratitude to the river. Make your amends once again. If possible, include your family or friends in your ritual process. Reach out to the river with your soul. Allow the river to speak to you. It will. To hear it all you need to do is empty your mind and just receive.

Is your special place in nature a bayou? Visit the bayou. Is it a lake? A pond? Wherever it is, sit quietly. Give yourself plenty of time. Pray and meditate. Absorb the beauty. Realize that it is disappearing forever. Allow yourself to feel sad. Go with a friend or family and share your memories and feelings. Again offer a prayer of gratitude to the body of water. Perhaps a song will come to you. Sing it.

Ask yourself how you can carry the message of Humanity Rising to other human beings. Remember, all this will be gone. Soon enough there will be no one to judge your actions, no one to broadcast disapproval at you. You can start up a local meeting of Humanity Rising. You can share with your friends on Facebook or another social media outlet. You can write a letter to the local paper.

Throughout, *Requiem* has mentioned basic 12-Step principles like honesty, open-mindedness, willingness, reverence, transformation, acceptance, kindness. How will you practice these principles in all you affairs? Can you be honest in your work? In your relationships with partner, children, family? Does work require dishonesty? How will you handle that? Can you honestly share your recovery with your friends and family?

Conclusion

There really is only one solution to the problem of the death of the world: The loggers will have to say, "No. We are not cutting down another tree." The soldiers will have to say, "No, we are not going to fight another war." The oil executives will have to say, "No, we will leave the oil in the ground." The commuters will have to say, "No, we are not driving to work or anywhere any more." The politicians will have to say, "No, we are not taking any more dark money from corporations." All of them will have to say, "Our planet, our lives, and our children's lives and the lives of all the wild animals and plants are worth more than any amount of money. We are creating a new way of living that protects and sustains the world and all life."

There really is only one solution.

www.ingramcontent.com/pod-product-compliance
Lightning Source LLC
Chambersburg PA
CBHW071305220526
45468CB00001B/272